MODELS OF SCIENCE TEACHER PREPARATION

Science & Technology Education Library

VOLUME 13

SCOPE

The book series *Science & Technology Education Library* provides a publication forum for scholarship in science and technology education. It aims to publish innovative books which are at the forefront of the field. Monographs as well as collections of papers will be published.

The titles published in this series are listed at the end of this volume.

Models of Science Teacher Preparation

Theory into Practice

Edited by

DERRICK R. LAVOIE

Virtual Institute for Teaching and Learning Science,
Paso Robles, CA, U.S.A.

and

WOLFF-MICHAEL ROTH

University of Victoria,
Victoria, BC, Canada

SPRINGER-SCIENCE+BUSINESS MEDIA, B.V.

A C.I.P. Catalogue record for this book is available from the Library of Congress.

ISBN 978-90-481-5755-6 ISBN 978-0-306-47230-5 (eBook)
DOI 10.1007/978-0-306-47230-5

Printed on acid-free paper

CONTENTS

ACKNOWLEDGMENTS

First and foremost, I wish to thank the chapter authors for outstanding perseverance through many drafts, internal chapter reviews, and discussions. I am particularly grateful for the instrumental support and assistance of co-editor, Michael Roth. I sincerely wish to thank those who provided critical reviews for the authors including: Ronald Anderson, Angela Barton, Ronald Bonnstetter, John Cannon, John Craven, Jim Ellis, Gerald Foster, Dorothy Gabel, James Gallagher, Jack Gerlovich, David Kumar, Norm Lederman, Cheryl Mason, Michael Matthews, Rena Norby, John Penick, and Larry Yore. I give special thanks to Ken Tobin who provided guidance and support throughout all phases of the project and to the reviewers of original outline and subsequent draft manuscript who helped us in shaping this book. Lastly, I dedicate this work to my family for their continual motivation and love during this venture.

Derrick Lavoie, April 2001

INTRODUCTION

Wolff Michael Roth[1] & Derrick R. Lavoie[2]
[1] University of Victoria, [2] Virtual Institute for Learning Resources

> The current reform in science education requires a substantive change in how science is taught. Implicit in this reform is an equally substantive change in professional development practices at all levels. (NRC, 1996, p. 56)

In a continuously changing society, it is not surprising that education also undergoes continuous change. Science education is no exception, and perhaps changes are more rapid given the daily construction of new scientific knowledge. In such a climate of continuous change, the preparation of science teachers has to follow suit in order to be appropriate to the reforms that national organizations encourage. However, whereas science teaching reform movements spawned recommendations of what teachers should know and be able to do in order for their students to conceptualize and process science (NSTA, 1997), they provide little guidance in terms of in-the-classroom concrete implementation. Thus,

> while national science education organizations continue to refine their positions about teacher education, there is no mechanism for translating these positions and statements into science education courses that can improve the preparation and quality of pre-service science teachers at both the elementary and secondary levels. (Yager & Penick, 1990, p. 670)

It is therefore not surprising that there are voices that describe teacher preparation as unsuccessful and as unresponsive to reform efforts (Schnur & Golby, 1995). Taking the high road, some science educators talk about new teachers in a negative way, that is, in terms of their 'resistance' to reform and to understand knowing and learning in constructivist ways. It is clear—studies such as the one reported by Roth (this volume) underscore this—that there is a gap between what new teachers hear in and experience at the university versus what they hear and experience "on the job". Rather than continuing the blaming cycle that new and practicing teachers are unresponsive to reform and changing theoretical frameworks, we need to find models for science teacher preparation that suitably assist prospective teachers to deal with the different experiences at the university and in schools. History provides evidence that past university-school relations are not good models for the future.

A cultural-historical analysis of teaching reveals that initially there was no separation between knowledge of teaching and knowledge about teaching (e.g., Roth & Lawless, in preparation). New teachers learned the ropes by being involved in teaching at the elbows of practicing teachers. In fact, this apprenticeship frequently began in the classroom when certain privileged students helped the regular teacher in accomplishing the task of teaching (e.g., Foucault, 1979). With industrialization and an increasing demand on a workforce that was educated and willing to submit to

D.R. Lavoie and W.-M. Roth (eds.), Models of Science Teacher Preparation, 1–8.

the work on the assembly line, schools began to adopt industrial models of administration ("scientific management"). This required, among other things, the training of teachers capable of "implementing" the curriculum decided upon by others and in the way these others deemed appropriate. The ensuing views on management gave rise to formal teacher training, which first was conducted in praxis-focused normal schools, but then was increasingly taken over by colleges and universities. That is, the division of labor between those who practiced teaching and those who generated the knowledge about teaching led to the well-known theory-practice gap, which, really should be understood as the politics of a theory-theory relationship in which academic theory is constructed as superior to praxis-based (often implicit) theory (Markard, 1997; Roth & Tobin, in press). Researchers increasingly distanced themselves from the experience of teaching in schools, taking an objectifying perspective on school events. Now, despite three or four decades of work on the theory-practice relation, (science) educators and particularly teachers continue to face a gap between the talk about practice by university-based science educators and the lived, concrete praxis of teaching science in schools. There surely is a need to revisit the relationship between universities and schools, and, correlatively, between theory and practice.

THEORY AND PRACTICE

Traditional View

Traditional ways of going about the university-school relationship are unsymmetrical. They put university and faculty members in the role of the knowing institutions that communicate this knowledge along a knowledge gradient to the less-knowing schools and teachers. Furthermore, such traditional ways assume that university-based researchers generate knowledge that is subsequently made available to teachers and schools. For example, the following statement discursively establishes such a gradient between research and the education of teachers.

> As ongoing and future research provides greater understanding of what constitutes constructivist teaching, researchers will need to address directly questions about how to educate teachers to successfully teach in this manner. It may well be the most needed—and potentially fruitful—area of research in science-teacher education. (Anderson & Mitchener, 1993, p. 26)

Here, *researchers* define and find out about "constructivist teaching" and subsequently "address questions about how to educate *teachers*". That is, in this and many other statements about the relationship between teaching, research, and teacher education, knowledge is made the prerogative of researchers. Teachers are relegated to the position of technicians who are supposed to do what they are being told.

Even the best-intended efforts to understand teaching and to influence the education of teachers fall into such traps. For example, Shulman (1987) devised a tripart model of teacher knowledge, which includes subject-matter content knowledge, pedagogical content knowledge, and general pedagogical knowledge. From the perspective of the university, this model underscored its role as expert and provider of this knowledge. Thus, science educators might interpret the Shulman model in the following way.

> For a science teacher preparation program to be effective it must harmoniously coordinate and integrate several components including a valid theoretical/research base, objectives, goals, content, process, assessment strategies, reflection and modification strategies, and student's cognitive, affective, and psychomotor attributes—all of which must be intimately linked with effective instructional models. For the science teacher, the Shulman model implies an integrated understanding of content as well as an understanding of student's learning styles and alternative scientific conceptions. For example, *Science for all Americans* is directed at integrating subject matter knowledge in the sciences. Curricular content knowledge centers upon goals, objectives, course design and development, course evaluation, and understanding ways to make the subject relevant to students daily lives.

In such statements, science educators demarcate the territory in which they are responsible for creating *valid* knowledge, as they are in control of the research base. Historically, knowledge of such things as objectives, assessment strategies, or content is the domain of university-based researchers.[1] They make it available to students, largely in lectures, and ask students to "construct" their knowledge. Such a tone is further reflected in statements that we can hear science educators make in their conversations about the need of science teachers. We might classify such talk—political correctness and review processes interfere with such talk making it into journals—as mere rhetoric. New and practicing teachers have their own way of describing such discourse as "talking the talk". It has been suggested that an understanding of learning in the social context of the workplace more generally can provide valuable information to the training of teachers more generally (Joyce & Clift, 1984; VerVelde, Horn, & Steinshouer, 1991). In our view, researchers need to move on and find different ways of working in schools, constructing relations with teachers, and understanding these relationships.

New Perspectives

Previous models for school-university relationships were not equitable, because one side in the division of labor (universities) constructed itself superior to the other (schools). In recent years, new more equitable models of interacting between universities and schools—often involving other partners as well—have been developed. These models are based on equitable cooperation in which the particular points of view of all partners are valued. For example, the reflective practicum (Schön, 1987) is an alternative to current practices in professional education which shares many features with traditional craft apprenticeship, athletics coaching, master classes in conservatories, and studios in architectural design. However, all too often, students are asked to reflect on the praxis of teaching, something that they have not yet experienced (Donnelly, 1999). Even many practicum, student teacher, and internship experiences are not authentic because new teachers are caught in the dilemma that they are not counted (in schools) as legitimate participants in teaching (e.g., Roth, Tobin, & Zimmermann, in press). They are still considered as "students", a status distinct from that of teacher, with too little experience to be taken for serious (e.g.,

Fetters and Vellom, p. 84, this volume). However, when the experience is authentic, newcomers to a practice learn by participating *with* old-timers in tasks that exhibit many characteristics of actual practice; that is, these tasks are *authentic*, but performed in a protected and safe studio-type learning environment. Such coparticipation is the core of the model presented in the first two chapters by Roth and by Thompson and Hargrave; such coparticipation could also be central to the problem-based learning approach presented here by Peterson and Treagust.

The issues at the core of these models chime with those of teacher educators who call for an ethos of learning in the workplace (VerVelde, Horn, & Steinshouer, 1991). They are also at the heart of models of professional practice schools that provide teacher-development activities around the notions of colleagueship, openness, and trust (Lieberman & Miller, 1990). Finally, the creation of school-based knowledge-building communities that include students, teachers, and teacher educators (Lieberman, 1992) is heading into the same direction. Such close and equitable relations between university and school communities are established in the models described in this volume by Fetters and Vellom and by Barufaldi and Reinhartz.

Until recently, forms of teacher education made few provisions for new teachers to develop their practice (Joyce & Clift, 1984). "Serious collaboration, by which teachers engage in the rigorous mutual examination of teaching and learning turns out to be rare" (Little, 1990, p. 187). Furthermore, little is known about the mechnisms by which collegial relations among teachers benefit student learning—though there exists at least one project in which such collegial relations have led to substantial science learning by students and teachers alike (Roth, 1998). There is certainly room for trying out and researching new forms of preparing new teachers for their full participation in the profession. It is in the spirit of change that we offer this edited volume to our readership.

CONTENTS

Collaboration and Apprenticeship

It is quite clear that at this stage, we need new and different approaches that navigate the line between theory and practice, teacher preparation and teaching praxis. Simply stated, these approaches need to be in stark contrast to the traditional "talking-the-talk"; university-based researchers and teacher educators need to begin "walking the walk". Collaboration between universities and schools, professors and (new and practicing) teachers in which none of the participants attempts to take undue control over the process of teaching, researching, and educating teachers show promise to become truly different from traditional approaches to the theory/research-praxis relationship.

The five models of teacher preparation presented in the first part of the book all focus in one way or another on collaboration and apprenticeship as models for preparing new teachers. Most models are also based on new forms of relationships between the different stakeholders involved in teaching and teacher preparation. Roth pushes the notion of collaboration. In his *coteaching*, new teachers work at the elbow of regular teachers, peers, university-based supervisors, methods teachers, and researchers. Here, the university-based participants no longer take the high road but roll up their sleeves and participate in getting the day's job done—teaching students science (e.g., Tobin, Roth, & Zimmermann, in press). All teachers and student representatives subsequently engage in cogenerative dialogue to generalize their understanding for the purpose of bringing about change that benefits student learning. In this model, learning to teach is but one aspect of the praxis of science teaching and benefits all participants in the relation. Their intention is to construct theory-in-practice, that is, to build conceptual knowledge through engagement in practice, by working at the school site (cf. Lieberman, 1992). Other forms of coparticipation are presented in the TEAMS model described by Thompson and Hargrave. Here, new teachers get to learn science and about science by shadowing scientists in their laboratories and teachers in the classrooms. Yet other models address the relationships between schools and universities at an institutional level (Fetters/Vellom and Barufaldi/Reinhartz). Whereas the former models are based on the engagement of individuals, these latter models lay the ground for sound collaborative efforts at a systemic level. These models therefore could serve as a macro organization for the earlier-described models that operate at a micro level.

Special-Issues Driven Models

The chapters in the second section of this book focus on special issues such as gender equity (Rennie), instructional congruence to promote science literacy (Lee & Fradd), portfolio assessment (Stein), new technologies (Lavoie), integrated-science curricula (Yager, Enger, & Guilbert), and critical multiculturalism (Thompson, Wilder, & Atwater). We clustered these models under the descriptor "special issues" because these models could be adapted to become part of one of the collaborative models presented in the first section. As recent research shows, reviewed by the respective authors, all of these issues have increased in their salience over the last decade. This also means that there is an increase in the number of issues that science educators are asked to address with the individuals in their charge, requiring us to rethink what and how we want to prepare new teachers. There may just not be sufficient time to teach all the pertinent issues in an explicit way. Again, allowing for different, implicit modes of learning to teach may be the answer to the apparent information bottleneck that all formal education has to grapple with.

Unified Chapter Focus

In this book, we offer a comprehensive number of chapters that focus on diverse issues that, as we have done, fall into two clusters. To bring some order into the expected diversity, we asked all chapter authors to follow, as much as possible, a common format. We thought that this would allow a better integration of complementary perspectives, frameworks, and purposes. Thus, each chapter endeavored to include the following components:

- Description of a model (or models) that focuses on a specific aspect of science teacher preparation. A "model" can be defined as an exemplary strategy employed to achieve a particular learning goal for science teacher preparation (e.g., incorporate technology, facilitate conceptual change, or enact gender equity).
- Description of a theoretical research-based perspective that specifically addresses the chapter's focus (i.e., the model) as well as reflects a larger encompassing constructivist paradigm for science education.
- Concrete descriptions of how the model can be bridged to practice (i.e., accomplished in the "real world") relative to existing science teacher preparation curriculum (e.g., goals, rationales, sequence, and design), coursework, and fieldwork. This will probably include vignettes and examples from the classroom.
- Descriptions of how the model or models can be assessed for its effectiveness in training science teachers as well as how the model can be modified. This may include a description of assessment practices that evaluate students' conceptual understanding, cognitive processing skill, psychomotor skills, science teaching attitudes as well as science teaching skills and knowledge. As well it may include descriptions of methods for (teacher) self-evaluation, student evaluation, program, and curriculum evaluation.
- Discussion of how the component model reflects the national goals and recommendations for the preparation of K–12 science teachers as well as national goals and recommendations for scientific understanding, attitudes, thinking skills, learning styles, and "science for all". Reference will be made to such documents as the *National Science Education Standards* (NRC, 1996), *Science and Engineering Indicators* (NSB, 1996), *Benchmarks for Science Literacy* (AAAS, 1993), and *Standards for Science Teacher Preparation* (NSTA, 1997).

We do not consider the models presented here to be prescriptive of any "ideal" program. Rather, the relationship between the different models presented here and the praxis of science teacher preparation is similar to the relationship between taking university-based science methods courses and the lived experience of teaching. The descriptions are *a priori* only potential resources for enacting science education and science teaching, respectively. Whether readers will have actually copied and implemented one or the other model described here is a matter of a posteriori analysis. We thought that science educators might want to use this volume as a resource to modify and expand their present programs so as to construct a uniquely harmonious and integrated science teacher training experience. By taking a very practical "in-the-trenches" approach with detailed descriptions and examples of on-going science teaching preparation practices, each model is potentially transferable to other science education programs. Readers are certainly encouraged to contact each author (team) to find out more about the specifics of enacting the model described.

This book is intended to appeal to science educators involved in the training of K–12 science teachers for inservice or preservice science teacher preparation and to graduate students in our discipline. Each chapter provides useful background, frameworks, guidelines, and concrete examples of program practices to science teacher educators seeking to enrich, innovate, and modify existing science education programs. The theoretical framework established in each chapter provides a research base for science education researchers and a resource to help grasp the relationship of theory and practice. The book timely attends to the *Standards for Science Teacher Preparation* and the *National Science Education Standards* and therefore should facilitate greater enactment of these standards in both schools and universities as well as stimulate insight for their revision. Finally, administrators, policy makers, and science education departments should find the book useful for course evaluation and revision procedures mandated by NCATE and other accrediting agencies.

NOTES

The writing of this introduction was made possible in part by a grant from the Social Sciences and Humanities Research Council of Canada (to Roth).

[1] At this point, we forego a discussion of the theories of learning implicit in the practices of stating objectives or the specification of learning content irrespective of student interest and associated assessment strategies. Suffice it to say that in this implicit theory, there is no room for a vision of individuals as co-creators of their life conditions. The theory is one of individuals as externally determined by their conditions. It therefore comes as no surprise that current practices of schooling reproduce existing inequities, including those along the line of gender, race, culture, and socioeconomic status (e.g., Barton, in press; Tobin, Seiler, & Walls, 1999).

REFERENCES

American Association for the Advancement of Science. (1993). *Benchmarks for science literacy: Project 2061*. New York: Oxford University Press.

Anderson, R. D., & Mitchener, C. P. (1994). Research on science teacher education. In D. L. Gabel (Ed.), *Handbook of research on science teaching and learning* (pp. 3-44). New York: Macmillan.

Barton, A. (in press). Science education in urban settings: Seeking new ways of praxis through critical ethnography. *Journal of Research in Science Teaching*.

Donnelly, J. F. (1999). Schooling Heidegger: On being in teaching. *Teaching and Teacher Education, 15*, 933-949.

Foucault, M. (1979). *Discipline and punish: The birth of the prison*. New York: Vintage Books.

Joyce, B. R., & Clift, R. T. (1984). Teacher education and the social context of the workplace. *Childhood Education, 61*, 115-119, 122-128.

Lieberman, A. (1992). Commentary: Pushing up from below: Changing schools and universities. *Teachers College Record, 93*, 717-724.

Lieberman, A., & Miller, L. (1990). Teacher development in professional practice schools. *Teachers College Record, 92*, 105-122.

Little, J. W. (1990). Teachers as colleagues. In A. Lieberman (Ed.), *Schools as collaborative cultures: Creating the future now* (pp. 165-193). New York: Falmer.

Markard, M. (1997, April). *Gramsci und psychologische Praxis oder: Psychologische Praxis als Austragungsort ideologischer Konflikte* [Gramsci and psychological praxis or: Psychological praxis as arena for ideological conflicts]. Paper presented at the Gramsci conference of InkriT. Available at: http://www.glasnost.de/autoren/markard/gramsci.html [2000, 09, 11]

National Research Council. (1996). *National science education standards*. Washington, D.C.: National Academy Press.

National Science Board. (1996). *Science and engineering indicators, 1996*. Washington, D. C.: U.S. Government Printing Office.

National Science Teachers Association. (1997). *NSTA standards for science teacher preparation*. Washington, D.C.: Author.

Roth, W.-M. (1998). *Designing communities*. Dordrecht, Netherlands: Kluwer Academic Publishing.

Roth, W.-M., & Lawless, D. V. (in preparation). *Being and becoming in the classroom*.

Roth, W.-M., & Tobin, K. (in press). Learning to teach science as practice. *Teaching and Teacher Education*.

Roth, W.-M., Tobin, K., & Zimmermann, A. (in press). Coteaching: Learning environments research as aspect of classroom praxis. *Learning Environments Research*.

Schnur, J. O., & Golby, M. J. (1995). Teacher education: A university mission. *Journal of Teacher Education, 46*, 11-18.

Shulman, L. S. (1987). Knowledge and teaching: Foundations of the new reform. *Harvard Educational Review, 57*, 1-22.

Tobin, K., Roth, W.-M., & Zimmermann, A. (in press). Learning to teach in urban schools. *Journal of Research in Science Teaching.*

Tobin, K., Seiler, G., & Walls, E. (1999). Reproduction of social class in the teaching and learning of science in urban high schools. *Research in Science Education, 29*, 171-187.

Vervelde, P., Horn, P., & Steinshouer, E. (1991). Teacher education: On site, on target. *Educational Leadership, 49*(3), 18-20.

Yager, R. E., & Penick, J. E. (1990). Science teacher education. In J. Goodlad (Ed.), *Teachers for our nation's schools*. San Francisco: Jossey-Bass Publishers.

SECTION ONE

COLLABORATION AND APPRENTICESHIP MODELS

BECOMING-IN-THE-CLASSROOM: LEARNING TO TEACH IN/AS PRAXIS

Wolff-Michael Roth
University of Victoria

INTRODUCTION

In teaching as in other domains, practitioners often make important distinctions between theories of their practice—which they encounter in formal settings (e.g., university course, seminar, or workshop)—and their experience of engaging in this practice in the here-and-now of the work place. In schools, I often hear teachers make distinctions between activities and teaching strategies that "only work in theory" or "should develop theoretically this way" and that "work differently in practice". That is, practicing teachers feel a gap between their lived experience in the classroom and the theoretical discourses that are supposed to explain this experience. The following comments—from teachers coteaching in K–12 classrooms who have very different experiences, backgrounds, and styles—illustrate this gap:

> The beginning was hard, for I remember at the university you're hearing all these ways and methods and these idealistic ways. But when you actually get out there in the schools it's different, when you try putting it into actions. I was sort of stumbling through things myself and it was a real struggle and a real battle. What I didn't really have was any modeling to follow. [Nadely, 4-month teaching intern]
>
> I just improved so much in teaching kids to think for themselves by asking productive questions. I don't think three university courses could have given me what coteaching gave me in these two months. [Tammy, 12-year veteran teacher]
>
> Even for someone who knows the unit theoretically, it is one thing to read it, but it is a whole different thing to do the unit with you. This coteaching experience has changed my thinking about this unit although I wrote it, tested it, and had done workshops with teachers on it for the past three years. [Gitte, curriculum developer and 4-year teacher]

All three were in the process, or at the end, of coteaching a science unit. As part of the experience, these teachers used videotapes when they wanted to observe their teaching and met with me and other teachers to talk about their science unit and their coteaching experience. These teachers—as all the other in the six coteaching projects I have conducted so far—acknowledged the tremendous learning trajectories they had followed and compared them to traditional modes of learning in 'acquire-theory → apply-in-practice' frameworks characteristic of teacher education. Unless we want to discount and marginalize the lived experiences of teachers, we need to take serious this experience of the difference between theory and practice and attempt to revise teacher education models to bridge this gap between university discourses and lived classroom experiences.

In this chapter, I present coteaching as a model of science teacher preparation. According to this model, individuals learn to teach science as part of praxis, while

D.R. Lavoie and W.-M. Roth (eds.), Models of Science Teacher Preparation, 11–30.

doing the day's work of a teacher. In coteaching, teachers learn essential elements of their practice that cannot be acquired through explicit modes of instruction. Coteaching also affords opportunities for learning through thinking-in-action that does not interfere with teaching itself. That is, coteaching is not about getting teachers trained in universities and then have them work with someone else in a classroom. Rather, coteaching is about substantially changing our ideas about science teacher education, supervision, evaluation, and research (Roth, Lawless, & Masciotra, in press-a, in press-b; Roth, Masciotra, & Boyd, 1999). Coteaching is not a quick fix, nor old wine in new bottles, but a way of situating learning to teach science in the praxis of teaching science.

PRACTICE, IN THEORY

In this section, I review the theoretical grounds on which our understanding of practice is based. The concepts used have their origin in practice theories (e.g., Bourdieu, 1980; Lave, 1988; Marx & Engels, 1970) and phenomenology (e.g., Heidegger, 1977; Merleau-Ponty, 1945).

From Theory to Practice

Traditionally, educational practice built on the findings of laboratory studies in domains such as cognitive psychology, problem-solving research, sociology, or artificial intelligence. Theories of teaching and learning and curriculum models developed from such research were then communicated to teachers who were supposed to implement what they had learned outside of the context of their work. This approach has largely failed to change the practice of teaching, and has frequently led to the negative attitudes practitioners voice regarding the "ivory tower", university professors' lack of understanding of educational practice, and the distance between theoretical discourses (characteristic of universities) and practical discourses of the work place.

In recent years, it has become increasingly clear to researchers in various fields that how humans think and act in laboratory studies is quite different from their thinking and acting in everyday settings. Their actual praxis has little to do with the formal description of their practices. Furthermore, talk *about* practice is distinctly different from practice itself (Bourdieu & Wacquant, 1992; Lave, 1988). For this reason, many researchers have left the confines of the laboratory to study ways of knowing in everyday settings without presuming traditional models of human rationality. In this way, investigators from various domains such as cognitive anthropology, sociology of scientific knowledge, or ethnomethodology have begun to investigate the cognition of scientists at their workbench. The participants include mathematicians and artificial intelligence researchers in front of their white boards; street bookies, grocery shoppers, and child street vendors in the street and supermarket; medical doctors at hospital beds; or navigation teams piloting their boats through narrow channels out toward sea (Cicourel, 1990; Hutchins, 1993; Lave, 1993; Traweek, 1988). This and other research has debunked the myth of human activity as a rational pursuit of goals by application of rule-based knowledge and suggests that human cognition is fundamentally situated in and distributed across specific social and material settings.

Most human activities are thus better understood in terms of practices, systems of structured actions that undergo specific, cultural-historical developments. These practices are very different from community to community, each of which is characterized by the conventions, standards, behavior, or viewpoints that its members (practitioners) share. In many work settings, people appropriate these practices when they coparticipate with others on the job. That is, in these situations, learning is not dissociated from the purposeful activity of getting the day's work done. The situation is different in teaching, for most (all?) preservice teachers never get to work side by side with a more experienced practitioner. Rather, they observe someone else teach or are largely left alone doing teaching. If the experience of learning in praxis in domains other than teaching is transportable, we may expect considerable gains in learning, and a lessening of the (traumatic) experience of the gap between *talking about* teaching (i.e., theory) and the lived *experience of* teaching (i.e., praxis). Before proposing a different model of preservice and inservice teacher education, I need to elaborate the nature of practical knowledge and mastery and the correlate concepts of "knowledgeability" and "spielraum". I also need to introduce being-with (Mitsein [Heidegger, 1977]), the fundamental condition of our experience that grounds (Donnelly, 1999) my concept of coteaching (Roth & Tobin, in press).

Masterful Practice as Knowledgeability

Master teachers generally know what to do based on mature and practiced understanding developed in the praxis of teaching. That is, they have evolved a knowledgeability, a sense for doing the right thing at the right time; in other words, they have developed a sense for acting "tactfully" (van Manen, 1995) and judiciously in situations where there is no time out for reflecting.[1] That is, in everyday teaching, there are many situations where the stop-and-think attitude characteristic of reflection can get teachers into trouble because they are out of synchrony with the unfolding events. However, master teachers do not need to stop and think; their thinking lies in (tactful and judicious) action that precedes all reflection (Masciotra & Roth, 1999; Roth, Lawless, & Masciotra, in press-b). Masterful practitioners are deeply involved in their work environment. They do not see problems in some detached way and take all the time to work at solving them (characteristic of the ways in which problems are solved in the ivory tower). They do not worry about the future and devise detailed and formalized plans as theoreticians would. Rather, masterful practitioners enter in a relation of interiority with the situation at hand, that is, they engage without the distancing inherent in a reflective, theoretical gaze (Dreyfus & Dreyfus, 1986; Roth, Masciotra, & Boyd, 1999).[2] Mastery entails fluid performance: we do not choose words nor place feet nor think how to think, we simply talk, walk, and think. Becoming a masterful practitioner is therefore a progression from (a) the analytic behavior of the detached subject in an objectified world, constituted by identifiable parts to (b) the involved skilled behavior based on accumulated experiences in particular settings and the preconscious recognition of new situations similar to previously experienced ones (Lave, 1996). Because our fundamental condition unavoidably is that of being-in-the-world, with physical bodies always situated in material, social, cultural, historical, and economic contexts, the process of developing practical mastery in teaching is therefore equivalent to becoming-in-the-classroom. But, because we always and already come to a "me"-predating world shot through with meaning, our being-in-the-world is always an experience of being-with (others).

In (sociological) phenomenology, being-in-the-world is fundamentally an experience of *being-with*. Even our experience of Self is secondary to the experience of symbiotic fusion with our mothers (Harré & Gillet, 1994) who are embedded in the patterns of their cultures and societies. Thinking is possible only once we can experience a separation of Self from Other, for the very underpinning of using representations (speech, tools, language, etc.) presupposes the use of something for something else (Ricœur, 1990). Thus, the social pervades us into the deepest categories that we use to structure our experience of Self and Other, subject and object, practical and theoretical knowing, and so forth (Bourdieu, 1980; Holzkamp, 1991). It is out of this experience of *being-with* that we nonthematically experience the ways others act in and towards the world and that we can experience their dispositions at work as they structure perceptions and actions.

In the phenomenological and practice literatures, there already exists a tradition of thinking about practice as something that precedes reflection. To theorize masterful practice, I draw on Heidegger's notion of *spielraum* (room to maneuver) which arises directly from the personal experience of being-in-the-world of practice. *Spielraum* is constituted by the "range of possibilities that Dasein 'knows' without reflection, [which] sets up *the room for maneuver* in the current situation" (Dreyfus, 1991, p. 190). Thus, when the situation does not allow for stopping-and-thinking but calls for immediate action, beginning practitioners often feel at a loss. The world of their experience contracts and becomes so small that they have no *spielraum* left, no room to maneuver. They freeze or might do something that they subsequently regret. In contrast, masterful practitioners, because of their knowledgeability in practice, can act *without* reflecting and tactfully engage in judicious action. In their experience of the situation, they have lots of *spielraum*. As practitioners become familiar with a context, their reality expands, and with it, their *spielraum*, and therefore the possibilities for acting. Thus, the existential possibilities open in any *particular* situation can be thought of as a subset of the *general* possibilities making up significance; and in this way, we achieve a definition of expertise that spans any particular situation in which it might find itself. (Such a shift between concrete possibilities and generalized possibilities available at the level of the society is also the heart of practice theories developed in Critical Psychology [e.g., Holzkamp, 1991].) These possibilities reveal what constitutes sensible, that is, tactful action in a specific situation. Masterful practice, when viewed in terms of knowledgeability and *spielraum*, limits what we might consider optimal environments for learning to teach. That is, it excludes those learning environments in which science teachers in preparation are asked to extract themselves from the situation at hand and to engage in (theory-oriented) distancing reflection.

KNOWLEDGEABILITY AND COTEACHING

Masterful teaching and the lived world of teaching differ from traditional theories of teaching that provide rationalized explanations for why teachers do what they do. Knowledgeability and *spielraum* are aspects of teaching that are intimately tied to the lifeworld of the classroom. This poses a dilemma for teacher education. If important aspects of teaching are ineffable, they cannot be brought into the discourse and therefore cannot be taught in university lecture halls. How then can teacher education contribute to the development of young teachers in these respects? My answer is, "Through coteaching!" Developing *spielraum* means that (beginning and in-service) teachers open up and increase the range of options for being and acting in the classroom. Such changes require participation in practice because it is only in practice that characteristic dispositions that generate perceptions and actions are enacted. Furthermore, because practice always operates under practical constraints of the unfolding events, practical decision making is different from theoretical pondering that does not have the same constraints. Thus,

> [t]here is no manner of mastering the fundamental principles of a practice ... than by practicing it alongside a kind of guide or coach who provides assurance and reassurance, who sets an example and who corrects you by putting forth, *in situation*, precepts applied directly to the *particular case* at hand. (Bourdieu, 1992, p. 221)

Thus, the modus operandi functions in a practical state, according to the norms of the trade and without necessitating that these norms be made explicit. It is a feel for what is right that causes us to do what we do at the right moment without needing to thematize (i.e., engaging in reflection) what had to be done and still less the knowledge of the explicit rule that allows us to describe this formidable practice. That is, coteaching allows teachers to experience the classroom at the elbows of another practitioner and thereby develop a sense of the practice through the eyes of the other. Through coteaching, teachers develop a sense of practice they share, a common sense. Bourdieu (1997) notes that such harmonization arises when people, with their material and social bodies, inhabit the same space and conditions. This affords an implicit collusion, a ratification and legitimization of a common practice ("everyone else is doing it", "ça se fait"). Learning to teach in and as part of everyday praxis allows beginning teachers to develop a sense of being a teacher; learning to teach is a process of becoming-in-the-classroom (Roth, Masciotra, & Boyd, 1999).

The interesting thing about the coteaching model is that the theory-practice gap does not exist. Teachers learn in practice and thereby acquire the modus operandi of teaching. They develop a habitus characteristic of teaching, ways of looking at (visions) and categorizing (di-visions) their lived world. Here, theory is implicit in practice, it is a "sens pratique", so that understanding is no longer separate from judicious action in a constantly changing classroom.

THEORY, IN PRACTICE

At the time of this writing, I have conducted 6 studies of coteaching in science classrooms totaling over 200 one-hour lessons each recorded by two cameras. My research team (including teachers) engaged in debriefing after each lesson and met regularly for two-hour sessions to talk about teaching and learning in coteaching situations. (At the time of this writing, I am conducting yet another study of coteaching in an urban school and as part of a teacher education program that has fully implemented the coteaching model [e.g., Roth & Tobin, in press].) In five of these studies, I was one of the two teachers coteaching science units with elementary teachers of different backgrounds and experiences. Gitte cotaught with the regular teacher in the sixth study (Roth, 1998b). Gitte, who split her time between developing curricula and graduate school, had taught full-time in grade 4 classrooms for four years and had developed curricula for another four years, including the unit on engineering design. It is out of these experiences and the endless conversations with teachers that I developed the notion of praxeology (Gr. *praxis*, action & *logos*, talk, speech), a praxis-situated way of understanding teaching to replace theories of teaching. We developed praxeology to replace the traditional methods and methodological instructions that are fraught with theoretical and practical problems (Roth, Lawless, & Tobin, in press).

School Participants

In order to assess the degree to which the coteaching model may be transportable to other settings (e.g., large scale and funded teacher inservice, preservice teacher education), I briefly describe the participants and settings in which the studies were conducted. All studies were conducted at the elementary level (K–7) in two schools in British Columbia. Mountain Elementary School offers in part a French immersion program and draws largely on a middle-class population in a large urban setting. There, I personally cotaught with Josephine (11 weeks) and with Cam (13 weeks). Seaside Middle School[3] is located in a semi-rural area with a considerable number of students from lower income homes and from the nearby First Nations communities. In each coteaching pair, there was one teacher with considerably subject matter expertise (Gitte or myself). The partners differed in their backgrounds and experiences. Cam succeeded Josephine as vice principal at Mountain Elementary; both had taught for more than 20 years, though it was Cam's first year at the elementary level. I cotaught a unit on simple machines with Josephine in her Grade 7 French immersion class and with Cam in his split Grade 6–7 class. Gitte cotaught with Tammy, a 12-year veteran at the Grade 4–5 level. At Seaside, I cotaught with both Nadely (14 weeks) and Loretta (8, 13 weeks) units on ecology including doing research in a local creek. Nadely had obtained an undergraduate degree in child and youth care and was in the process of completing the capstone four-month internship of a two-year professional program. Loretta had taught for nearly 15 years at the Grade 7 level and was the only classroom teacher with a science background (B.Sc.).

The invitation of the research team to participate in the school's activities did not go without trepidation in both schools. There were initial ambiguities about the researchers' roles and initial anxieties about having a camera in the classroom that had been observed in other university-school efforts (e.g., Clark et al., 1996; Clift, Johnson, Holland, & Veal, 1992). However, in both schools there were teachers (i.e., those featured here) who took the lead and invited me to their classrooms to participate in science teaching and subsequently invited me and my team to work with them and their classes. Research team and cameras became an integral part of the classroom and data collection disappeared into the background.

Coteaching, in Practice

In my work on coteaching, the outside teacher became as much a part of school culture as possible. Both Gitte and I participated in full-day professional development activities, staff meetings, and curriculum planning sessions with the entire staff. At Mountain Elementary, the staff had voted to make the improvement of science teaching the primary ongoing concern; at Seaside, half a dozen teachers, supported by their principal, wanted to work with me on improving their science teaching. In addition to coteaching one unit in a class, Gitte and I both also worked concurrently with teachers of other classes, conducting individual lessons, planned curriculum with them, cotaught their classes, and met for post mortem sessions so that we could learn from what had happened. Thus, our roles were not those of external experts who came to the schools to present "dog-and-pony shows", though these had been requested at times. Rather, Gitte and I always emphasized that our contributions to school life would lead to learning only if we worked together *with* resident teachers on all aspects of the lessons.

Gitte and I each planned the activities together with the respective teacher to adapt them to the specific needs of the setting. When requested, I provided support by supplying videotapes for individual or group viewing, serving as sounding board for reflection after the lessons. More crucially, the teachers and visiting teacher-researchers shared stories of their own learning with the children so that the classroom became a "learning community" (Roth, 1998a). Visits by other teachers, which were explained to students as occasions for the visitors to learn about teaching and learning, heightened the general sense of the classroom as a learning community.

The model of teachers who work together is not new. Cooperation among teachers has been shown to be beneficial and to lead to teacher learning at their work place (Joyce & Clift, 1984). Team-teaching has been a concept and practice for some time (Ollman, 1992). I prefer the notion of coteaching to emphasize the doing-together and being-with the other rather than the teaching (in parallel) in the team situation—though there are situations in coteaching were both teachers will go off and work individually with different student groups. By talking about coteaching, I want to emphasize that there are situations in which one teacher literally works at the elbows of the other, thereby experiencing the classroom from the perspective of the other. This is of particular importance, for in order to understand actions, we need to know the world through the eyes of the actor (Bakhtin, 1993; Markard,

2000), we need to know her *spielraum*. Coteaching then allows one of the teachers to take time out during small moments to stop and think, to reflect on what is happening while it is happening. Whereas this would not be a good move for the teacher currently taking the major responsibility—because it interferes with fluid performance (Masciotra & Roth, 1999; Roth, Lawless, & Masciotra, in press)—it permits the other to bring to awareness some situation and therefore develop new understandings.[4] Such reflection, which is almost contemporaneous with the action, allows the teacher to bring to awareness relevant precepts as they are applied to the *particular case* at hand. The second mode of learning in coparticipation involves the appropriation of practices not through becoming aware of their structure, but because of the tendency to do what everyone else does, or because this is the way to do and act in the case at hand. This learning is akin to the Aristotelian notion of *mimesis* (Roth, 1998c) and finds an explanation in the Heideggerian understanding that we always come to a world already shot through with meaning (Roth, 1997).

Reflecting-on Reflection-in-Action

Coparticipating allows for stopping-and-thinking, reflection during the activity of teaching with only a minor delay between what is happening and the reflection (which presupposes, and requires time for, re-presentation). Our debriefing and reflection-on-action sessions (where we often used videotape to transport us back to the situation) allowed us to get a sense for our respective experiences in particular teaching sequences. In the following excerpt, I had begun a whole-class session to deal with the question one student (Carl) had raised during the previous lesson. Carl had wondered what would happen if electricity was conducted through water. I had brought a light bulb, battery, and a few wires connected into a circuit that was open at one point. I first touched the two open ends, thereby lighting up the bulb. Subsequently, I stuck the two leads into a beaker with water. The episode picks up after students noted that nothing had happened. At the same time, Nadely stepped a bit back and experienced the situation with a reflective attitude (as retold during our reflection-on-action session).

Michael: Nothing, OK, so, there is no light coming on, this is amazing because they... Aren't they telling us on TV that water is conductive?
Bill: But we need to add salt!
Jon: May be it's the wire?
Michael: OK, someone said salt, someone talked about salt. (Bran: I did) What do you think happens if I put that [Holds two leads] in the salt here?
Bill: Well, nothing
Stan: You got to add the water also.
Arlene: Try it
Michael: OK, we'll try it. [Puts 2 leads in salt] Nothing is happening.
Bill: You need salt in the water for the whole, it'd give you more of a chem... I think it'll.

[Nadely: I was listening to your questions and how you've formatted that question at that moment. And that helped me then and there. Because it triggers something in my head, I go, 'Oh yeah I should be thinking about that', or 'I should be asking about that', or 'That was a good question that really got them going on this tangent or brought them back or got them more focused.'

Carl: We'll have a chemical reaction.
Michael: OK, What happens if? What happens if we had some of the salt in here [Points to beaker with water]?
Bill: Mix it, mix it around!
Michael: We have to, we have to mix it. [Mixes salt in beaker] What happens if I put the wires in there *now*?
Tory: It'll work.
Stan: Chemical reaction!
Tory: It'll work.
Michael: Why would it? Explain! Who has an explanation?

Yeah, that was helpful for me to, to listen to your questions and hear your questions then and there, *as it happened, and then to think about how it related to the demonstration and where you were trying to go with that question.*

When I reflected with Nadely on this episode, I became aware that I did not think about students' conceptions, about teaching strategies, or about subject matter pedagogy. Rather, I recalled that I was simply a presence, ready to evolve a conversation with the students whatever the previous answer. Much like in everyday conversation, I talked rather than choosing words and planning to talk. My experience is best described as a sense of heightened awareness. It is a sense of being-in and comprehending the situation without having to thematize it, that is, to objectify it in the various ways I am able to objectify physical settings, teaching, students, relationships, etc. This heightened awareness experience was not unlike that which I had lived as a high-performance athlete 20 years earlier. Here too, I was ready to deal with the contingencies arising during a race without being able to predict what these contingencies would be (gale, wave by motor boat, attack by one or another opponent). Thus, my questions were not preformatted, nor did I have prepared answers for I could not have known students' responses and questions. It was simply that I felt to have considerable *spielraum*, a background sense that I could act tactfully and judiciously however the conversation would evolve.

When Bill suggested that salt was needed to make the light come on, I spontaneously asked the question what would happen if the two open leads were pushed into the salt. Without reflecting, the student suggestion had opened an inquiry into the question whether salt is conductive. Of course, after the fact I am able to rationalize my actions. Salt water is conductive; the conductivity comes from the dissociation of salt molecule NaCl into the two ions Na^+ and Cl^-. The salt molecule itself is not conductive, and therefore I do not expect that the light comes on. Furthermore, students have already seen that water did not conduct electricity (at least not well enough for the light to come on) because only one tenth of one millionth of water molecules H_2O are dissociated, that is, in ionic form (H^+, OH^-) which makes them conductive. Therefore, there was an opportunity that students would experience a cognitive conflict that makes them think about the question why neither salt nor water are conductive but a mixture of the two is. However, while I can rationalize my actions after the fact this does not mean that they were causally predetermined by these rationalizations. Rather, it is more likely that I embody the structures of the chemistry and physics. Having been "disciplined" into the two subjects during my graduate studies, I embody the structures of these disciplines (cf., Bourdieu, 1997; Roth, 1999) so that I can look at the world and see phenomena *as instances of* Newtonian forces, chemical reactions, heat transfer, or entropy.[5] This embodiment allows me to find myself at home without deliberation whatever students suggested. Without thinking about it, I was able to enact *the* questions to ask; and I produced, 'comme il faut', questions that existed as objective possibilities that oriented my practice.

Nadely's relation to the questions was different. Having time out from leading the whole-class conversation, she could reflect, but without loosing track of the ongoing activity. That is, she reflected as the activity evolved. All reflection, however, requires an object, a re-presentation, which is different from the thing it stands for. Representation can only occur once she had extracted herself from the ongoing action. Yet she reflected on the action without, nevertheless, having the extended time required for reflection-on-action. That is, this reflection was in activity, and yet distinct from action.

When we sat together watching the video clip of such episodes, we did have the time to truly reflect-on-action.[6] Here, Nadely recited some of the patterns in questioning that she had become aware of during her reflection-in-action. For example, she noticed that I asked students to explain, justify, or elaborate on what they had previously said. (See also the chapter by Hargrave and Thomspon, p. 40.)

> [Nadely:] I was just accepting yes or no as answers, but there were no explanations for them. So I know that when *you* would get up there and form your questions in a little bit, slightly different way that would ask the students to give more than just a one-word answer, that it would trigger things for me.

Whereas she allowed students to get away with one-word and yes-no answers, my questioning opened up opportunities for longer conversations and contributions. But constructing and bringing to awareness some explicit rules about questioning gives rise only to a symbolic mastery of a practice (Bourdieu, 1980). There is often a long way from realizing how we should teach and actually teaching accordingly (Roth, 1998b); and there is another step before we can achieve practical mastery (Masciotra & Roth, 1999). However, using practical rules even if they cannot express the nature of excellence, allows us to understand differences between our own actions and those of others. They therefore allow us to monitor our actions:

> [Nadely:] When I got up there and asked a question, I began to realize, 'Oh no that wasn't good enough.' I could hear myself asking the close-ended questions.

We see here that Nadely has learned in the sense that she not only notices differences between her own and my questions when observing herself on video but she also begins to notice the differences as she was asking them. However, this is not the only or even principle mode of learning through coparticipation. There is also a mode of learning were we do what others do without ever becoming aware that we are doing so.

"I Do as You Do": Pedagogy of Silence

Coparticipation affords learning in practice without bringing the practices into awareness. Whereas some science educators may find this a strange idea, there are many situations in our everyday lives where we learn in this mode. For example, children learn to produce well-formed sentences in their mother tongue without ever receiving formal instruction in grammar. More so, even without instruction people are able to distinguish correct from incorrect sentences although they may have never heard these before, and without ever having had one grammar lesson. The lesson we can learn from this is that we can develop expertise without having descriptions of it.[7] My research shows that coteaching allows for such learning and that one teacher comes to make her own the questions and ways of talking subject matter she had experienced with her partner.

Gitte was a teacher with great subject-matter expertise who was recognized for her practical mastery of questioning (Roth, 1996, 1998b). As she cotaught with Tammy, the two began to resemble each other in striking ways. My notes on videotapes, field notes, and my annotations of transcripts reflected this sense. I frequently recorded remarks such as, "Tammy is asking a question, but I hear Gitte" or "Gitte is doing what Tammy would have done". As they worked together, Gitte and Tammy began to resemble each other in their mannerisms, their overbearing stance toward some students and a more supportive stance towards others. The videotapes show how an individual, pensive movement of a hand to the chin by Gitte was followed almost instantaneously by the same motion of the hand by Tammy; a turn of the head or the whole body of Tammy was reflected in the movements of Gitte. Here, Tammy and Gitte developed common ways of acting and manners of judging without that they ever talked about it or brought it into reflective awareness.

After about two months, the two were so attuned in their practices that they could conduct a class without previously orchestrating their roles.

> [Gitte:] I thought that was really neat. Well this morning when we were questioning, did you notice how I asked the kids to answer and then you were...? I noticed this we were standing next to each other, and were looking out of the corner of our eyes, and we just sort of knew to alternate like that without having spoken about it.

Gitte and Tammy performed questioning because they "just knew to alternate like that without having spoken about it". This accord, the often tacit understanding of the coparticipant, attested to the fact that they had learned more from each other than they could state in just so many words. They developed an accord that does not presuppose communication (and therefore interpretation) or contractual decisions and which gives rise to a practice-oriented "intercomprehension" (Bourdieu, 1997). Such an accord arises out of the experience of *being-with,* which allows that practices are "objectively harmonized without any calculation or conscious reference to a norm and mutually adjusted *in the absence of any direct interaction* or, *a forteriori*, explicit coordination" (Bourdieu, 1980, p. 98, my translation, italics in the original).

One area of interest in my research was the nature of Tammy's questioning and how it might have changed during the coteaching experiences. Here, the videotapes document changes in Tammy's use productive questions. Within a short period of time, she had included several productive questions into each interaction with students. However, what the videotapes did not reveal was how this change of practice was brought about, that is, how Tammy had learned. While watching the videotapes of this lesson, Tammy had become aware of differences between her and Gitte's questioning. She had come to realize that her way of asking questions was not productive; that is, she "was going nowhere". But upon reflection, she was impressed with the way Gitte brought students to think about the content matter through her questioning. This suggests that she brought about changes in a conscious process. However, she also indicated that she would not have used the questions had she read them in the teachers' manual. Although this manual explicitly elaborated relevant content and pedagogical content knowledge, Tammy found it difficult and time-intensive to learn from the manual. That is, although the manual was written by a teacher for teachers in a non-technical manner, including many illustrations and examples, Tammy did not peruse it. It was somehow too far removed from the experience of actually teaching the unit with Gitte. Experiencing the question in the context of teaching the engineering unit "put it into her being" and "made it become part of her". Tammy emphasized that participating with Gitte helped her to make the questions part of herself, something she would not have been able to had she read the same questions in a resource book. Hearing questions in context helped her in the appropriation of the practice which, in an unexplained manner, "goes in", "makes it part of" her so that she can use it.

Practical knowledge does not arise from a relation of exteriority that is characteristic of a conscious knowing. Rather, practical knowledge arises as a matter of course from the bodily presence of human agents in the material and social world which they inhabit (Bourdieu, 1997). This bodily presence—which allows for, and is constitutive of society ("corps sociale")—affords pre-reflective appropriation, mimesis. Just as our physical behavior is structured by gravity—our main body axis is parallel to the gravitational force—so our social behavior is structured by the patterns of society, and particularly communities of practice, that embed us. Here, the two teachers appropriated behaviors from each other, in a pedagogy of silence, without attending to what and how they learned (changed their practices). That is, even without explicit intentions to change, coteaching allows for the homogenization of teaching practices that arises from the experience of *being-with*, that is, from coparticipating in the same classroom and building a history of common experiences, themselves forming the individual habitus of each partner.

EVALUATION

The praxis of coteaching is grounded in the phenomenological idea that understanding presupposes *being-with* another and sharing the same place from which to view the world. Evaluation of coteaching therefore requires the "evaluator" to participate in coteaching (Roth & Tobin, 1999). That is, coteaching presupposes that faculty supervisors, teacher evaluators, teacher trainers, and so on also participate in coteaching because this is the only position from which they can properly understand why a (preservice) teacher does what she does. At the University of Pennsylvania, (science) teacher preparation, supervision, and evaluation is being implemented through coteaching (Tobin, Seiler, & Smith, 1999). Being-with a teacher, sharing in planning, teaching, debriefing, evaluation, assessment, or supervising is what coteaching is about. Praxeology, enacted as conversations about the common teaching experiences (which may draw on reflection-on-action, but also on phenomenological accounts of experience), allows teachers to understand themselves as well as their coteaching partners. It is out of this understanding developed in shared democratic conversations without traditional power relationships that come co-authored evaluation reports, suggestions for new strategies, supervision reports, and research understandings.

The advantages of teacher-teacher collaboration, coteaching as a mechanism for achieving job performance, for teacher development and reproduction of the teacher community is virtually nonexistent (Joyce & Clift, 1984; Lieberman & Miller, 1990). However, my six studies all showed the positive conditions for change when practitioners coteach. More so, our recent work in a teacher education program shows the tremendous potential of coteaching as a means to prepare future teachers (Roth & Tobin, 1999, in press). Coteaching not only led to observable changes documented in my extensive video records (e.g., Masciotra & Roth, 1999; Roth, 1996, 1998a, 1998b, 1998c; Roth, Bowen, Boyd, & Boutonné, 1998; Roth & Boyd, 1999; Roth, Masciotra, & Boyd, 1999), coteaching was above all experienced by teachers as something positive.

> [Tammy:] After watching Gitte, I just improved dramatically. I realized that I was going nowhere fast, and I wasn't helping these kids at all with the kinds of questions. I could have done this unit without Gitte, but I would never be where I'm right now, with those kids. The kids would have never made the bridges they made today, because I just wasn't able to think enough. To me the whole process of organizing them and getting them to do what I was expecting them to do took so much of my thinking. I wouldn't have had the time to think about the questioning, and that was so much more important, that would bring out so much more.

In this quotation, Tammy highlights the challenges of teaching a new, more child-centered unit on engineering which were mediated in her work with Gitte. Rather than having to focus explicitly on questioning, adding to her planning and preparation, she learned questioning in the course of teaching the unit. The quotes in the introductory section already showed how participating teachers felt that coteaching provided for learning opportunities that were both intensive and yet seemed to be more efficient than what traditional university courses could provide them with. Nadely further elaborates this point:

> [Nadely:] The university provides you with examples or scenarios, and maybe what you can do, ideas. But really where you learn is the *here and now*, because every group is so different and every individual that's in that group. This class is the way this class is because of the individuals that are in there.... I learned a lot from you: Now this happened to me today and this is how I dealt with it and it didn't work or this happened today and it did work.

Nadely recognized that university courses can only provide scenarios, descriptions sufficiently abstract that they fit many situations, but in this, not any individual one. Where the real learning occurs, then, is in the here and now of the classroom. While she cotaught with me, she personally experienced ways of acting and interacting, and had time to reflect in action when I took over, because she could step back and thereby remove herself from the urgency that unfolding events demand from the person in charge. In the process, Nadely developed *spielraum* to engage in a range of actions without having to stop-and-think. (For a detailed description of this development see Roth, Masciotra, and Boyd, 1999.) This development was precipitated both implicitly, a pedagogy of silence, and explicitly, through reflection-on-action as the lesson unfolded and in post mortem sessions.

In our everyday work as teachers, we face many situations where taking time out for reflection is not an option. Nevertheless our practical mastery allows us to act tactfully and judiciously. However, some readers may find themselves agreeing with the critic in the following metalogue. (From anonymous reviewer comments to another article [Roth, Masciotra, & Boyd, 1999].)

METALOGUE

Critical reader: I have no problem with your claim about the importance of tacit dimensions of teaching and being in the classroom. But I am unconvinced that your performance, and hence the learning of Nadely or Tammy, was not an example of the very things—reflection-in-action (Schön, 1987) and pedagogical content knowledge (Shulman, 1987)—you wish us to believe the coteaching goes beyond.

Michael: Although you seem to be sympathetic to my argument, you also construct only rational ways of knowing as legitimate. However, if we admit only rational forms of knowledge, we marginalize and thereby deny our own lived experiences and therefore teaching as praxis. That is, the very experience our theories are to explain would be deleted from professional discourse.

Critical reader: My problem is that you, as the expert coteacher, know about pedagogical content knowledge as an intellectual concept, not just in the sense of innately incorporated subject teaching knowledge. You are very familiar with reflection on practice. So you are not really going beyond these in your teaching.

Michael: I admit to be familiar with and engage in these discourses. But there is also a difference in post mortem discourse about questioning and the experience of questioning in situ with no time out. Furthermore, the fact that I have a discourse to describe and explain my actions does not imply that this discourse causally precedes my actions. Ethnomethodological researchers have attempted to bring exactly this precarious relationship between plans (descriptions) and actions.

Bourdieu: In my way of framing these issues, Michael's discourse only gives him symbolic mastery over the practices. However, his practical sense enacts questioning either without representing them or while giving itself only partial or inadequate representations of them. Furthermore, excellence (that is, practical mastery in its accomplished form) ceases to exist as soon as we ask whether it can be taught, as soon as we seek to base canonical practice on rules extracted, for the purposes of transmission, from the practices of previous periods.[8]

Critical reader: Well, there is still the problem of generalizability of coteaching to other settings. Without evidence of some other regular teacher being equally effective in developing Nadely, Tammy, and others through coteaching, your model will remain quite unconvincing to most hard-pressed supervising teachers in schools. It also will be unconvincing to many teacher educators who will not feel able on a regular basis to act with their teachers as you did so admirably.

Michael: For one, coteaching allowed Tammy to develop expertise in questioning, although Gitte did not have my background with regard to reflection-in-action and reflection-on-action, or with regard to the three categories of pedagogical, pedagogical content, and subject matter knowledge. Our recent experience of coteaching as a way of preparing urban teachers, we found that your so-called 'hard-pressed supervising teachers' thoroughly enjoyed coteaching, learned a lot from it, and were seeking opportunities for further coteaching experiences.[9]

Bourdieu: I think we can get around the problems if we abdicate the notion that all behaviors are rule-driven. I believe that rules spuriously occupy the place of two fundamental notions, theoretical model and practical sense, and therefore prevent us to question the relationship between the two. Although a rational pedagogy may use a theoretical model to serve practical functions, we have to recognize that theory is completely alien to practice.[10]

Angie: Let me press the issue. Following your earlier argument, a beginning teacher experiences a theory-practice gap. Why would this change in co-teaching situations? Perhaps this has to do with how one conceptualizes what that gap is. For example, what if the beginning teacher and co-teacher do not share similar world-views in terms of subject matter knowledge, schools, or society?[11]

Michael: I agree that this would be a difficult issue. Ken Tobin and I talked a lot about his own experience of teaching with someone else who does not share his commitments. This person also chooses to move out of the field of action rather than in coteaching with Ken so that both teachers and the students benefit from this experience. Such incompatibilities can become constraints to what can be learned by all people involved in the situation.

Angie: I also wondered what it was about the coteaching and the learning done by Nadely that was dependent on coteaching? Could she have had the same reflections about your questioning sequence if she was team teaching or observing?

Michael: Again, my conversations with Ken Tobin suggest that we gain very different perceptions of teaching practices when we are actually looking at it from the inside, literally from a point of view like that of the other teacher. Details of what it means to observe teaching from the inside, and what differences there are, if any, between inside-out and outside-in perspectives, still have to be worked out through appropriately designed research studies.

DISCUSSION AND IMPLICATIONS

From my six studies on coteaching, and from its correlative theoretical framework, emerges a view of teacher knowledge that contrasts traditional views of science teachers' cognition. Traditional, cognitive perspectives highlight the formalizable knowledge of a domain, that which is expressible as declarative and procedural knowledge (Grimmett & MacKinnon, 1992). These theories presuppose that once apprehended, knowledge is applied in concrete, problematic situations in an applied science sense. Classical teacher training programs built on this view provide future teachers with subject matter knowledge and pedagogical knowledge. These programs are based on "the traditional shibboleths of 'theory into practice' or 'application of knowledge'" (Lieberman, 1992, p. 718). It is assumed that all teachers have to do is apply this knowledge in unproblematic ways in their professional settings. Time and again, beginning teachers feel deceived and frustrated with this approach because they experience little overlap between "ivory tower" theories and practice as they experience it, between the "idealistic ways" and the ways of everyday praxis. My work breaks with such (traditional) conceptions of theory and practice, by locating a central aspect of learning into praxis, and thereby takes serious the by teachers experienced gap between theory and practice. (Recently, Ken Tobin and I have collaborated in working out some of the implications of such a view with respect to the practices of teacher development, supervision, evaluation, and research [Roth & Tobin, in press].) Praxis implies both getting the job done (teaching kids) and learning to teach (professional development). Marxist thinkers also tell us that this is the only way that theory can truly be tested, *in* practice and *for* practice (Marx & Engels, 1970).

Our praxeology, talk (Gr. *logos*, talk) about praxis (Gr. *praxis,* action) from the perspective of practice (Roth & Tobin, 1999), is in sharp contrast with classical views of practice. It is consistent with recent practice theories according to which only those who participate in praxis can know it (e.g., Dreier, 1993). Accordingly, to learn the essence of a practice, newcomers have to coparticipate in it. However, coparticipation means that teachers coteach, work at each other's elbows, not merely teach in the same school or be in the same classroom. *Being-with* implies 'being-in-the-same shoes', doing the same things, and looking at the world from similar social locations (e.g., as teachers in the classroom). Learning arises from increasing coparticipation in competent practice and practice-related discourse. Newcomers appropriate ways of structuring their experience in community-specific ways, that is, they develop a habitus. Coteaching also provides common conditions that serve newcomers as referents for reflection with their partner. They therefore appropriate more formal aspects of knowing (rules) and become more competent in linking formal, explicit and practice-constitutive, tacit knowledge.

The model is consistent with recent reform movement in science and mathematics education that focus on students' acquisition of authentic scientific and mathematical practices (NCTM, 1989; NRC, 1996). My work extends this focus on the learning of science and mathematics to the pedagogical content knowledge appropriated in these fields by newcomers and old-timers alike when they coparticipate directly with others in the workplace. Furthermore, the coteaching model is consistent with reform documents that foster bottom up rather than top–down-type changes. The emphasis in bottom up approaches is on professional development rather than on administrative directives. Thus, inservice teacher programs "must include mechanisms for sustained collegial interaction, links between staff development and classroom practice, and the participation of administrators to ensure support for the proposed changes" (NCTM, 1989, p. 253). Ken Tobin and I, by participating in the day-to-day work of teaching science in public schools, have radicalized the notion of "support". Furthermore, the NCTM guidelines ask for conditions that coteaching has already built in given that the necessary resources [shared experiences], time to reflect, and opportunities to share ideas with other teachers. However, as with all design changes, coteaching demands a transformation of (preservice and inservice) teacher education models and thus involve the introduction of "new goals, structures, and roles that transform familiar ways of doing things into new ways of solving persistent problems" (Cuban, in NCTM, 1989, p. 251).

Since I have started my work, the teacher education program at the University of Pennsylvania under the leadership of Ken Tobin has implemented coteaching across the board (e.g., Roth & Tobin, in press; Tobin, Seiler, & Smith, 1999; Tobin, 2000). The initial results, though there are problems, are highly promising. Even beginning teachers who initially negotiated to teach on their own later moved on their own into coteaching relationships with peers and experienced teachers. The coteaching model may also turn out to be a cost-effective alternative to the traditional summer workshop approach for providing inservice teacher education. Simple calculations show that if school boards began to have a few "master teachers" (in the true sense of the word as it evolved from apprenticeship models in craft domains, but with a dual direction of learning as described by Schön [1987]) who spend several months with each teacher, no more than three at any one time, the costs of inservice (in the US mostly funded through NSF and State Departments of Education) could be dramatically decreased.

Coteaching should also be feasible for preservice teacher education. Most teacher education programs have already established school-university relationships (on this relationship, see also the chapters by Barufaldi/Reinhartz and Fetters/Vellom). However, two questions need to be seriously addressed. First, "What are the criteria that an experienced teacher would have to meet to be acceptable as a coteacher?" Second, "How can we help mentor teachers change from their current sink-or-swim model to a coteaching model?" Here, I do not have ready answers. In my own studies, both Gitte and I had considerable understanding of the subject matter content and the associated authentic scientific and engineering practices. We were committed to science education as an endeavor that engages children in cultural production and reproduction. Furthermore, our institutional constraints were of a different nature than those that practicing teachers experience as part of their work. Both of us were also committed to coteaching as an activity that affords learning to all participants in the classroom community. Ultimately, those most concerned in each partnership situation need to make the choices and work out feasible ways of making coteaching work, in situation.

Finally, coteaching provides for a new model of teacher supervision, evaluation, and research. Thus, as we know from personal experience (e.g., Roth & Tobin, 1999), looking at acts of teaching from the outside is very different depending on one's perspective. Thus, we are presently exploring coteaching as a new way of enacting teacher supervision, evaluation, and research and we are testing the insider perspectives against those established from the outside—sitting on the sidelines in the back of the classroom (e.g., Roth & Tobin, in preparation).

EPILOGUE

In this chapter, I presented our experiences with coteaching as a model of science teacher preparation. As I have outlined it here, coteaching requires a radical break with traditional notions of what makes a competent science teacher. Coteaching requires among others that faculty walk the walk about science teaching and coparticipate in the daily work of teaching children. Being-there and being-with another teacher allows them to see teaching as it is, what possibilities for action particular moments in classroom life give rise to and so on. Being-there in the classroom and being-with another teacher and students is the primary place for understandings to emerge. This requires commitment to change that will take different trajectories depending on where you currently are.

NOTES

Data collection and writing was made possible in part by Grants 410-93-1127, 812-93-0006, and 410-96-0681 from the Social Sciences and Humanities Research Council of Canada. My thanks go to those who, in various ways, participated in my studies of coteaching: Nadely Boyd, Sylvie Boutonné, Michael Bowen, Fiona Crofton, Josette Désquin, Allan MacKinnon, Michelle McGinn, Ken Neale, Laurie Roche, Christina Schnetzler, Bridget Walshe, and five classes of students. My thanks are extended to Domenico Masciotra who helped me to think through important issue related to the experience of teaching in real time. I want to thank Ken Tobin, Gayle Seiler, and Maxwell Smith for sharing their own experiences with coteaching and changes in science teacher preparation during extensive discussions. Last but not least, I thank Angie Barton, who was a tough critique and a helpful consultant for this chapter.

[1] I judiciously chose the word "sense" to link to Bourdieu's (1980) French title of "Le sens pratique", practical sense, which has been translated as "The logic of practice". However, sens pratique affords Bourdieu a link to "sens objectivé" (objectified sense) and sense commun (common sense).

[2] Reflection presupposes an object that can be reflected upon, and therefore a withdrawal from and objectification of the situation. In this, reflection presupposes time out.

[3] This school district includes grades 6 and 7, which in other parts of British Columbia are elementary school grades, in their middle school program (grades 6–8).

[4] This is consistent with the analysis that understanding arises from the dialectic between phenomenological experience and a reflexive, hermeneutic analysis (Ricœur, 1991). Similarly, the "sens pratique" [practical sense] in dialectical opposition of "sens objectivé" [objectified sense] gives rise to a "sens commun" [common sense] (Bourdieu, 1980).

[5] As I have shown elsewhere, students do not embody these structures and therefore *cannot* see in laboratory activities or demonstration the science that teachers expect them to (e.g., Roth, McRobbie, Lucas, & Boutonné, 1997a, 1997b).

[6] According to Schön (1987), reflection-on-action is an activity in which some previous action is made the object of thought; for this to happen, the individual has to extract herself from action and objectify it before she can consider it within the context of other objectified actions. Schön constitutes reflection-in-action as a form of problem solving in real time of action. For a critique of the notions of reflection-in-action to describe teacher know how see Roth, Lawless, and Masciotra (in press-a).

[7] Incidentally, connectionist approaches to modeling intelligence and cognitive behavior come to the same result. That is, neural networks learn from examples and learn to respond intelligently without ever learning explicit rules that describe these responses (e.g., Churchland, 1995).

[8] Adapted from Bourdieu, 1980, p. 174f.

[9] Cf. Roth & Tobin, in preparation.

[10] Adapted from Bourdieu, 1980, p. 176.

[11] Adapted from a private communication to me by Angie Barton (February 10, 1999).

REFERENCES

Bakhtin, M. M. (1993). *Toward a philosophy of the act.* Austin: University of Texas Press.

Bourdieu, P. (1980). *Le sens pratique* [The logic of practice]. Paris: Les Éditions de Minuit. [Translation available from Polity Press, Cambridge, 1990]

Bourdieu, P. (1992). The practice of reflexive sociology (The Paris workshop). In P. Bourdieu & L. J. D. Wacquant, *An invitation to reflexive sociology*. Chicago, IL: The University of Chicago Press.

Bourdieu, P., & Wacquant, L. J. D. (1992). *An invitation to reflexive sociology*. Chicago, IL: The University of Chicago Press.

Bourdieu, P. (1997). *Méditations pascaliennes* [Pascalian meditations]. Paris: Seuil.

Churchland, P. M. (1995). *The engine of reason, the seat of the soul: A philosophical journey into the brain*. Cambridge, Mass: MIT.

Cicourel, A.V. (1990). The integration of distributed knowledge in collaborative medical diagnosis. In J. Galegher, R. E. Kraut, & C. Egido (Eds.), *Intellectual teamwork: Social and technological foundations of cooperative work* (pp. 221-242). Hillsdale, NJ: Lawrence Erlbaum Associates.

Clark, C., Moss, P. A., Goering, S., Herter, R. J., Lamar, B., Leonard, D., Robbins, S., Russell, M., Templin, M., & Wascha, K. (1996). Collaboration as dialogue: Teachers and researchers engaged in conversation and professional development. *American Educational Research Journal, 33*, 193-231.

Clift, R., Johnson, M., Holland, P., & Veal, M. L. (1992). Developing the potential for collaborative school leadership. *American Educational Research Journal, 29*, 877-908.

Donnelly, J. (1999). Schooling Heidegger: On being in teaching. *Teaching and Teacher Education, 15*, 933-949.

Dreier, O. (1993). Re-searching psychotherapeutic practice. In S. Chaiklin & J. Lave (Eds.), *Understanding practice: Perspectives on activity and context* (pp. 105-124). Cambridge: Cambridge University Press.

Dreyfus, H. L. (1991). *Being-in-the-world: A commentary on Heidegger's 'Being and Time', division I*. Cambridge, Mass: MIT.

Dreyfus, H. L., & Dreyfus, S. E. (1986). *Mind over machine: The power of human intuition and expertise in the era of the computer*. New York: Free Press.

Grimmett, P. P., & MacKinnon, A. M. (1992). Craft knowlege and the education of teachers. *Review of Research in Education, 18*, 385-456.

Harré, R., & Gillett, G. (1994). *The discursive mind*. Thousand Oaks, CA: Sage.

Heidegger, M. (1977). *Sein und Zeit* [Being and time]. Tübingen, Germany: Max Niemeyer. (English translation consulted is that by Joan Stambaugh, State University of New York Press, 1996.)

Holzkamp, K. (1991). Societal and individual life processes. In C. W. Tolman & W. Maiers (eds), *Critical psychology: Contributions to an historical science of the subject* (pp. 50-64). Cambridge: Cambridge University Press.

Hutchins, E. (1993). Learning to navigate. In S. Chaiklin & J. Lave (Eds.), *Understanding practice: Perspectives on activity and context* (pp. 35-63). Cambridge: Cambridge University Press.

Joyce, B. R., & Clift, R. T. (1984). Teacher education and the social context of the workplace. *Childhood Education, 61*, 115-119, 122-128.

Lave, J. (1988). *Cognition in practice: Mind, mathematics and culture in everyday life*. Cambridge: Cambridge University Press.

Lave, J. (1993). The practice of learning. In S. Chaiklin & J. Lave (Eds.), *Understanding practice: Perspectives on activity and context* (pp. 3-32). Cambridge: Cambridge University Press.

Lave, J. (1996). Teaching, as learning, in practice. *Mind, Culture, and Activity, 3*, 149-164.

Lieberman, A. (1992). Commentary: Pushing up from below: Changing schools and universities. *Teachers College Record, 93*, 717-724.

Lieberman, A., & Miller, L. (1990). Teacher development in professional practice schools. *Teachers College Record, 92*, 105-122.

Markard, M. (2000, June). Kritische Psychologie: Methodik vom Standpunkt des Subjekts [Critical Psychology: Methodology from the perspective of the subject]. *Forum Qualitative Sozialforschung / Forum Qualitative Social Science [On-line Journal], 1*(2). Available at http://qualitative-research.net/fqs/fqs-d/2-00inhalt-d.htm [access: August 15, 2000].

Marx, K., & Engels, F. (1970). *The German ideology* (C. J. Arthur, Ed.; W. Lough, C. Dutt, & C. P. Magill, Trans.). New York: International.

Masciotra, D., & Roth, W.-M. (1999, March). *Beyond reflection-in-action: A case study of questioning in science teaching*. Paper presented at the annual conference of the National Association for Research in Science Teaching, Boston, Mass.

Merleau-Ponty, M. (1945). *Phénoménologie de la perception* [Phenomenology of perception]. Paris: Gallimard.
National Council of Teachers of Mathematics. (1989). *Curriculum and evaluation standards for school mathematics*. Reston, VA: NCTM.
National Research Council. (1996). *National science education standards*. Washington: National Academy Press.
Ollman, H. E. (1992). Team teaching with a student teacher (Open for suggestions). *Journal of Reading, 35*, 656-657.
Ricœur, P. (1990). *Soi-même comme un autre* [Oneself as another]. Paris: Seuil. (Published in English by Chicago University Press, 1992.)
Ricœur, P. (1991). *From text to action: Essays in hermeneutics, II*. Evanston, IL: Northwestern University Press.
Roth, W.-M. (1996). Teacher questioning in an open-inquiry learning environment: Interactions of context, content, and student responses. *Journal of Research in Science Teaching, 33*, 709-736.
Roth, W.-M. (1997). Being-in-the-world and the horizons of learning: Heidegger, Wittgenstein and cognition. *Interchange, 28*, 145-157.
Roth, W.-M. (1998a). *Designing communities*. Dordrecht, Netherlands: Kluwer Academic Publishing.
Roth, W.-M. (1998b). Science teaching as knowledgeability: A case study of knowing and learning during coteaching. *Science Education, 82*, 357-377.
Roth, W.-M. (1998c). Teaching and learning as everyday activity. In K. Tobin & B. Fraser (Ed.), *International handbook of science education* (pp. 169-181). Dordrecht, Netherlands: Kluwer Academic Publishing.
Roth, W.-M. (1999). Discourse and agency in school science laboratories. *Discourse Processes, 28*, 27-60.
Roth, W.-M., Bowen, G. M., Boyd, N., & Boutonné, S. (1998). Coparticipation as mode for learning to teach science. In S. L. Gibbons & J. O. Anderson (Eds.), *Connections 98* (pp. 80-88). Victoria, BC: University of Victoria.
Roth, W.-M., & Boyd, N. (1999). Coteaching, as colearning, in practice. *Research in Science Education, 29*, 51-67.
Roth, W.-M., Lawless, D., & Tobin, K. (in press). Towards a praxeology of teaching. *Canadian Journal of Education*.
Roth, W.-M., Lawless, D., & Masciotra, D. (in press-a). Relationality as an alternative to reflectivity. *Journal of Teacher Education*.
Roth, W.-M., Lawless, D., & Masciotra, D. (in press-b). Spielraum and teaching. *Curriculum Inquiry*.
Roth, W.-M., Masciotra, D., & Boyd, N. (1999). Becoming-in-the-classroom: A case study of teacher development through coteaching. *Teaching and Teacher Education, 17*, 771-784.
Roth, W.-M., McRobbie, C., Lucas, K. B., & Boutonné, S. (1997a). The local production of order in traditional science laboratories: A phenomenological analysis. *Learning and Instruction, 7*, 107-136.
Roth, W.-M., McRobbie, C., Lucas, K. B., & Boutonné, S. (1997b). Why do students fail to learn from demonstrations? A social practice perspective on learning in physics. *Journal of Research in Science Teaching, 34*, 509-533.
Roth, W.-M., & Tobin, K. (1999). Toward an epistemology of teaching as practice. Submitted.
Roth, W.-M., & Tobin, K. (in press). Learning to teach science as praxis. *Teaching and Teacher Education*.
Roth, W.-M., & Tobin, K. (in preparation). *Learning to teach in urban schools*. New York: Peter Lang.
Schön, D. A. (1987). *Educating the reflective practitioner*. San Francisco: Jossey-Bass.
Shulman, L. S. (1987). Knowledge and teaching: Foundations of the new reform. *Harvard Educational Review, 57*, 1-22.
Tobin, K. (2000). Becoming an urban science educator. *Research in Science Education, 30*, 89-106.
Tobin, K., Seiler, G., & Smith, M. W. (1999). Educating science teachers for the sociocultural diversity of urban schools. *Research in Science Education, 29*, 68-88
Traweek, S. (1988). *Beamtimes and lifetimes: The world of high energy physicists*. Cambridge, MA: MIT Press.
van Manen, M. (1995). On the epistemology of reflective practice. *Teachers and Teaching: Theory and Practice, 1*, 33-50.

TEAMS: A SCIENCE LEARNING AND TEACHING APPRENTICESHIP MODEL

Constance P. Hargrave & Ann D. Thompson
Iowa State University

INTRODUCTION

In this chapter, we present a model for enhancing the science preparation of preservice teachers by creating science learning and teaching apprenticeships. Based on cognitive apprenticeship theory (Collins, Brown, & Newman, 1989), our science learning and teaching apprenticeship model is grounded in experiences that make the pedagogical expertise of classroom teachers and the science process and content expertise of scientists visible for preservice teachers. (See also the chapter by Roth in this issue.) Collaborations among scientists, practicing teachers, preservice teachers, and teacher educators were part of the Teacher Education and Achievement in Mathematics and Science (TEAMS) project in which our model was tested.

COGNITIVE APPRENTICESHIP THEORY

Science teaching preparation of elementary teachers typically consists of a series of introductory science content courses (e.g. introductory biology, environmental geology, and general chemistry) and a science teaching methods course (Goodlad, 1990). The science content courses are often taken independently of the science teaching methods course and are taught by science content experts who have little or no formal preparation as educators (Raizen & Michelsohn, 1994). As a result, the science learning experiences of preservice teachers occur in isolation from their thinking about and preparation in science pedagogy. Teaching elementary science (like teaching any subject) is a dynamic and complex skill that requires integration of many types of knowledge (Roth, 1996). To meaningfully teach elementary science one must possess science content knowledge (AAAS, 1990; Grossman, Wilson, & Shulman, 1989), knowledge of instructional strategies (Radford, 1998), knowledge of children's intellectual and social development (Anderson & Mitchener, 1994), and skills for implementing all of these types of knowledge with groups of young people (Doyle, 1990; Trumbull, 1990).

Researchers in the areas of learning and cognition are giving increasing attention to the roles of situation and activity in learning (Greeno, Moore, & Smith, 1993) particularly with respect to learning to teach (Roth, 1998a). Concepts are not abstract, self-contained, and independent entities. Rather, conceptual knowledge is situated and progressively developed through authentic activity (Brown, Collins, &

D.R. Lavoie and W.-M. Roth (eds.), Models of Science Teacher Preparation, 31–47.

Duguid, 1989). Learning to teach science is inextricably linked to the context in which the learning occurs as well as the activities in which one engages (Roth, 1998b).

We view the process of becoming a science educator as one of enculturation (see also the chapter by Roth in this volume). That is, becoming a science educator requires more than obtaining discrete knowledge of science and pedagogy. Rather, the process of becoming a science educator requires immersion into the scientific and science education communities. Because scientists and science educators have their respective professional cultures, immersion must be such that the individual develops a conceptual and experiential understanding of the practices, norms, and language of the community. In this way, the science concepts and the science teaching strategies are situated in their naturally complex contexts.

Immersion is characterized by participation in activities of the culture (Brown et al., 1989). Consistent with this idea, the *National Science Education Standards* (NRC, 1996) call for an increase in authentic scientific inquiries conducted by students. Activities are authentic when they engage people in the ordinary practices of a culture. Putting these two ideas together suggests that we need to develop experiences and activities for preservice teachers that expose them to the ordinary practice of science.

Like any profession, the respective activities of scientists and science educators are defined by their culture. The purpose and meaning of a culture's activities are socially constructed through negotiations among members. Oftentimes the activities of a culture are communicated to newcomers through apprenticeship activities. The basic premise of an apprenticeship is to show a novice how a complex task is done and then teach the novice to do the task. Collins et al. (1989) outline four aspects of traditional apprenticeships that are applicable to cognitive apprenticeships: modeling, scaffolding, fading, and coaching. In *modeling*, the expert demonstrates the various phases of the task for the novice. In so doing, the expert makes each step of the phases visible for the novice. *Scaffolding* is the process by means of which the expert gives support to the novice in carrying out the task. This support can vary from extensive to minimal based on the knowledge and skill of the novice. *Fading* is the process of decreasing the level of support the expert provides and increasing the novice's level of responsibility in carrying out the task. *Coaching* is the essence of the apprenticeship; it is the overall guidance the expert provides the novice throughout the entire learning period. The expert defines the tasks, provides explicit and implicit instruction and direction, evaluates the novice's activities, diagnoses problems, and provides feedback, encouragement, and guidance. This expert therefore relates differently to the novice than in the coteaching example provided by Roth (this volume).

An extension of traditional apprenticeship, "the goal of cognitive apprenticeship is to make the thinking processes of a learning activity visible to both the students and the teacher" (Collins, Brown, & Holum, 1991, p. 39). The science learning and teaching apprenticeship was developed to bring together preservice teachers with experts in science content and science teaching in such a way as to permit them to observe, engage in, and discover experts' strategies in context (i.e. the science laboratory and the elementary science classroom). In the TEAMS model, our goal was to make the thinking processes of the scientists and the classroom teachers visible to the preservice teachers. Teacher educators, working with the scientists and classroom teachers, used the aspects of traditional apprenticeship (modeling, coaching, scaffolding, and fading) to develop the preservice teachers' skills in, and knowledge of, science teaching.

SCIENCE LEARNING AND TEACHING APPRENTICESHIPS THAT REFLECT NATIONAL STANDARDS FOR SCIENCE TEACHER PREPARATION

The science learning and teaching apprenticeship model described in this chapter reflects the emphasis on authentic science experiences in current work on reform in science education. In this section, connections of the TEAMS model to the major thrusts of national reform efforts in science education are summarized to provide context and rationale for the later description of TEAMS.

The current era of science curriculum reform emphasizes context-rich, in-depth science experiences for students and preservice teachers. National projects such as Science/Technology/Society, Project Scope, Sequence, and Coordination, Project 2061, and other interdisciplinary approaches to the teaching of science necessitate the development of new effective models of preservice science teacher education that emphasize authentic and connected science experiences for preservice teachers. *Revitalizing Teacher Preparation in Science—An Agenda for Action* (NSTA, 1993) provides useful insights connecting teacher education reform with reform in science education. The report draws from the reform literature in preservice teacher education and focuses it's six principles through the eyes of leading scientists, educators, and science educators. These principles suggest that every elementary-middle-secondary preservice science teacher should (a) experience the investigative nature of science; (b) have classroom and laboratory experiences in biology, chemistry, earth/space science, and physics; (c) understand the inter-relatedness of science disciplines and the connections between science and other areas of knowledge; (d) learn scientific content and thinking processes in the context of contemporary, relevant personal and societal issues and problems; (e) have a sound understanding of the nature of learning and how it can be applied to the learning of science; and (f) have several intense and extended clinical teaching experiences at a variety of grade levels in diverse socioeconomic and cultural settings.

The six guiding principles focus our attention on the interaction of content/process preparation, pedagogy, and the nature of human development and learning. Thoughtful teacher education efforts must focus on the intersections of these three areas of expertise. The TEAMS project is a practical model that attempts to provide an integrated science education experience for preservice students that helps create links among content, pedagogy and learning.

In November 1994, the National Research Council first published *Standards for the Professional Development of Teachers of Science*. These standards, built upon research and practice, serve as the basis for developing new models of science teacher education. These Professional Development Standards include standards for learning science content, learning to teach science, learning to learn and program development (NRC, 1996). It should be noted that several of the specific standards from this work relate directly to the design and implementation of the TEAMS model described in this chapter. These related standards include:

1. Science learning experiences involve teachers in actively investigating scientific phenomena, interpreting results, and making personal sense of findings consistent with currently accepted scientific understanding. (National Science Education Standards, 1996; Professional Development Standard A p. 60-61)
2. Learning experiences address issues, events, problems, or topics significant in science and of interest to participants. (NRC, 1996; Professional Development Standard A p. 61)
3. Teachers are introduced to scientific literature, media, and technological resources that expand their science knowledge and their ability to access further knowledge. (NRC, 1996; Professional Development Standard A p. 61)
4. Teachers are to be encouraged and supported in efforts to collaborate. (NRC, 1996; Professional Development Standard B p. 67)
5. Learning to teach science takes place in actual classrooms to illustrate and model effective science teaching and permit teachers to struggle with real situations and practice, and expand knowledge and skills in an appropriate context. (NRC, 1996; Professional Development Standard B p. 63, 67-68)
6. Opportunities are provided for teachers to receive feedback about their teaching and to understand, analyze, and apply that feedback to improve their practice. (NRC, 1996; Professional Development Standard C p. 68)
7. Preparing and using mentors, teacher advisors, coaches, lead teachers, and resource teachers to provide professional development opportunities supports sharing of teacher expertise. (NRC, 1996; Professional Development Standard D p. 71)
8. People involved in programs, including teachers, teacher educators, scientists, administrators, policymakers, and business people must collaborate—with clear respect for the unique perspectives and expertise of each. (NRC, 1996; Professional Development Standard D p. 71)

Designed to improve the quality of the mathematics and science preparation of elementary preservice teachers through the involvement of scientists and practicing elementary teachers, the TEAMS model builds upon the broad principles recommended by a variety of leading national organizations. Reflective of the national reform emphasis on authentic and context-rich science experiences for preservice teachers, the TEAMS model exemplifies many of the concepts discussed in Roth's chapter. Moreover, the TEAMS model focuses specifically upon putting into practice selected principles for science teacher preparation recommended by the National Research Council (1994).

SCIENCE LEARNING AND TEACHING APPRENTICESHIPS

The TEAMS model is a cooperative effort between scientists from the Ames Laboratory, Iowa State University (ISU) teacher educators, and outstanding elementary school teachers from the Ames community. Ames Laboratory, of the U.S. Department of Energy, is housed on the ISU campus and employs research scientists to study various scientific phenomena ranging from crystallography to metallurgy. The TEAMS model enables preservice elementary teachers to learn science content and teaching methods in authentic environments. Moreover, this model provides the opportunity for the preservice teachers, as well as the professionals involved, to discuss and reflect upon these experiences.

Participant Selection and Project Structure

Undergraduate education majors who have been accepted into the teacher education program (usually in the junior year) are considered for TEAMS based on mathematics/science experiences, grade point average, teacher recommendations, and standardized test scores. Generally, the applicants have completed basic education courses such as strategies in teaching, foundations of education, and instructional technology and are enrolled in discipline-specific teaching methods courses. Each preservice teacher selected for participation is placed on a squad that consists of three to five preservice teachers, one scientist, one classroom teacher, and one teacher educator. Generally, there are four squads each term. Working in squads, the preservice teachers spend two hours each week with their scientist in a research laboratory and two hours in the classroom of their practicing teacher. In addition, all preservice teachers meet in a weekly one-hour seminar with teacher educators, a scientist, and a teacher.

The Role of the Scientist and the Science Laboratory Experience

Under the direction of the scientist, the science laboratory experience provides the preservice teachers with opportunities to be a part of a science research team, experience the processes of science, and develop professional relationships with individuals in the scientific community. The scientists in the laboratory have three goals. First, they want to introduce the preservice teachers to scientific research by exposing them to the research conducted in the laboratory and providing a big picture context of the purpose and implications of the research (modeling). Second, they want to involve the preservice teachers in the processes of science by designing investigations that are subsets of the laboratory research yet appropriate for the knowledge and skill level of the preservice teachers (scaffolding and fading). Finally, they want to serve as science mentors to the preservice teachers by advising them about science content for their teaching experiences (coaching), as well as reflecting with the preservice teachers on the practice of science.

Shadowing and Hands-On Activities in the Science Laboratory

The laboratory component was planned as a shadowing experience for the preservice teachers. The preservice teachers were to develop their understanding of scientific research by being in a rich science environment, following the scientists around the laboratory, observing the tasks in which they engaged, and asking questions about science content and the implications of the studies. Early attempts at this approach proved unsuccessful; scientists reported that it was not manageable or meaningful. They stated that shadowing was too vague in that it did not outline specific activities in which the scientists could engage that would be meaningful for the preservice teachers to observe. In addition, the shadowing experience did not specify what the preservice teachers were to gain from their observations of the scientists. In the terms of the cognitive apprenticeship model, the scientists were modeling behaviors; however, the scaffolding necessary to make the modeling meaningful for the preservice teachers was missing.

The scientists explained that their activities in the laboratory were often too varied for the preservice teachers to learn much science. ("I do too much running around putting out fires; following me around would be a disaster. They wouldn't get anything out of it".) Furthermore, because of the highly specialized nature of the science research at Ames Laboratory, the scientists suggested that the students did not have sufficient science background, level of expertise, or vocabulary needed to make shadowing a meaningful experience. To accommodate the students' needs, the scientists, collectively and individually, redefined the laboratory component from a shadowing to a hands-on experience. It was decided that each scientist would develop and engage the preservice teachers in hands-on activities based on research in their laboratory. This modification of the original vision allowed the scientists to provide the scaffolding and eventually fading necessary for the novices to enter the world of the expert scientists.

For example, Dr. Scott Chumbley is a materials scientist who studies the microstructures of metals. He is specifically interested in the composition of metal microstructures, the mechanical properties of metals, and the effects of micro-structural changes on the properties of metals. His primary tools for examining microstructures and properties of metals are microscopes and various mechanical tests. Thus, the laboratory activities Dr. Chumbley developed for TEAMS involved two basic tasks: microscopy and mechanical testing. To acclimate students to the laboratory environment, Dr. Chumbley conducted tours of his research environment. In so doing, he provided the preservice teachers with a broad understanding of materials science research and the implications of his work. Through the use of the optical and electron scanning microscopes, the hands-on activities began with basic examinations of common metals. These included brass and steel nuts and aluminum from beverage containers. The activities progressed to include the examination of laboratory research samples, discussions about data analysis, operation of sample preparation equipment, examination of sample properties, and implementation of simple mechanical tests. Dr. Chumbley also worked with the students and their master teachers in the elementary school classroom. Together, the squad designed and then implemented classroom lessons based on the students' laboratory experiences. During the development and execution of the lessons, Dr. Chumbley and the classroom teacher served as coaches for the novice teachers.

The Role of the Classroom Teacher and the Classroom Experience

To increase the preservice teachers' exposure to and experience in the classroom, they spent two hours per week in the room of the practicing classroom teacher on their squad. (Each practicing teacher who participated in TEAMS was the recipient of the Presidential Award for Excellence in Mathematics or Science Teaching.) The purpose of the classroom experience was to provide the preservice teachers with an opportunity to critically analyze the processes and techniques of effective science teaching and to incorporate these methods into their own teaching practices and beliefs about education. The goals of the practicing teachers included: (a) modeling effective hands-on teaching of science for the novice teachers; (b) reflecting upon their expert teaching strategies for the novice teachers (scaffolding); (c) providing opportunities for novice teachers to teach independently in the classroom (fading); (d) evaluating the novice teachers work in the classroom (coaching); (e) facilitating the development of the preservice teachers' science instructional knowledge and skills; and (f) increasing the preservice teachers' confidence in their ability to effectively teach science.

The practicing teachers were those who would most influence the teaching philosophies and strategies of the preservice teachers in TEAMS. Thus, the primary objective of the practicing teachers was to model (with reflection) effective classroom practice. As a professional role model for the preservice teachers, the practicing teachers demonstrated (through their regular classroom practices) a variety of strategies that create authentic environments for science learning.

An essential component of the classroom experience was the guided reflection component that provided necessary scaffolding for the preservice teachers. As part of each week's experience, the practicing teacher would spend 15-20 minutes with the preservice teachers reflecting on the classroom activities that occurred. One teacher summed up her role by stating "My goal is to let them see what went right or wrong with my lessons". These guided reflection opportunities assisted the preservice teachers in analyzing the teaching and learning processes and helped them to better understand the techniques, strategies, and processes of effective science teaching. The scaffolding provided through the reflections with the classroom teachers provided guidance that the novice teachers found extremely useful.

Shadowing in the Classroom

Similar to the laboratory experience, the classroom component was originally envisioned to be one where the preservice teachers would "shadow" the teachers. That is, the preservice teachers were to observe the actions of the practicing teacher and participate in the classroom environment as directed by the teacher. The notion of a shadowing experience worked well in the classroom. Because of their academic background and knowledge of teaching, the guided-observation experience was meaningful for the preservice teachers. The modeling of the classroom teachers was much easier for the students to relate to than the initial modeling of the scientists.

Comments from the preservice teachers during interviews and focus groups suggested that the classroom component was a valuable experience. "Suzanne is so open. She shares what she is doing and what she is thinking," a preservice teacher commented. Although all of the preservice teachers prior to TEAMS had observed classrooms (as part of the teacher preparation program), they indicated that the classroom component of TEAMS was different than their other experiences. They reported that the teacher modeling in guided observations and the scaffolding provided in the time for reflection significantly helped them learn from the classroom teacher. ("I got a sense of what a really good teacher is like.... I got to look at teachers differently.... I got a sense of figuring out why they do what they do and ask what they ask".)

The Role of the Teacher Educator and the Weekly Seminar

Because the purpose of the TEAMS model is to improve preservice teachers' knowledge of science and skill in teaching science through collaborations with classroom teachers and scientists, the primary role of the teacher educators was to bring together professional communities that historically have not been closely connected. Thus, the teacher educators were the cohesive factor in TEAMS. The teacher educators served as translators, mediators, and facilitators for the preservice teachers' experiences in laboratory and classroom and in the preservice teachers' efforts to transfer their science learning into grade-level–appropriate science lessons. The teacher educators provided scaffolding to help bring together the laboratory and classroom experiences.

The creation of a community, in which everyone can be comfortable and productive, requires strong communication skills. The teacher educators were the professionals best situated to understand the professional perspectives of the other participants (i.e. scientists, classroom teachers, and preservice teachers) and, as such, facilitate communication between and among them.

The teacher educators fulfilled their roles in a manner that matched the personality of the squad. For example, to build communication bridges between the scientist and the preservice teachers on his squad, Dr. Gary Downs (a professor of science education) participated in the initial experiments the preservice teachers conducted with the scientist. His participation helped the scientist frame research problems in a manner that was meaningful for the preservice teachers. For example, the scientist on Dr. Downs' squad investigated the processing of powders to improve permanent magnets. Initially, the scientist began the TEAMS laboratory experience by exploring the properties of various metal powders. Aware of the preservice teachers' science background, Dr. Downs suggested that before exploring metal powders, the squad should examine how magnets function. With this prompting, the scientist and the preservice teachers experimented with magnets and cooperatively described their characteristics. As a result of this experience, the scientist better understood the preservice teachers' knowledge of magnets and metals and designed a series of small experiments to help the preservice teachers realize the role of permanent magnets and the need for metal powder processing. During the initial experiments, Dr. Downs asked many questions of the scientist (Why are some metals magnetic? how are magnets used in appliances? What is the difference between permanent and non-permanent magnets?) By asking questions, Dr. Downs helped the preservice teachers feel comfortable in what otherwise may have been an intimidating environment.

To facilitate communication with her squad, Dr. Chris Ohana (professor of science education) accompanied her squad's scientist to the elementary science classroom of the participating teacher where they watched a lesson and then met to plan TEAMS activities. After watching the lesson, the scientist indicated that he was overwhelmed by the task of teaching science to children and intimidated by the classroom teacher's ability to develop and implement meaningful science-learning experiences for 30 fifth graders. For the next planning meeting, Dr. Ohana's squad met in the laboratory. Before the meeting, the squad scientist conducted a tour of the laboratories, gave a general overview of the research being conducted, and introduced the squad to various scientists who were engaged in research. By walking through the facility, seeing the wide variety of individuals who participated in the research, and observing the interactions between scientists, the classroom teacher began to understand the nature of the research and the environment in which the scientist worked. These initial meetings, coordinated by the teacher educator, helped participants to identify connections between their respective professions that would assist the preservice teachers.

Although teacher educators fulfilled their roles in a unique manner, two activities were used with all TEAMS participants to encourage communication: weekly seminars and e-mail journals. The weekly seminars helped establish a community among all the participants and provided a natural context to bring together the science laboratory and classroom experiences. Facilitated by teacher educators, the weekly seminars provided a forum for the scientists to articulate their understanding of teaching, an opportunity for the teachers to articulate their conceptions of science, and a collegial environment for the preservice teachers to think about the tasks of learning and teaching science. (Weekly seminar topics included the cultures of science research and science teaching, classroom resources available for science instruction, multicultural and gender issues in science [see also chapters by Rennie and Thompson et al., respectively], constructivist science teaching, and squad progress reports.)

To further assist the preservice teachers in developing their conceptions about science instruction, e-mail journals were used. Each week in seminar, the preservice teachers were given questions to encourage them to reflect upon, and articulate their understandings and beliefs about, science and teaching science. A listserve was created for each squad, and the preservice teachers' journal entries went to each member of their squad, including the scientist, classroom teacher, and teacher educator. The other preservice teachers on the squad, as well as the professionals, would react and respond to each other's journal entries; this created an on-going, on-line dialog where the preservice teachers and the experts expanded and refined their individual and collective understandings of science and science teaching.

Assessment of the Preservice Teachers during TEAMS

In keeping with apprenticeship theory, assessment for TEAMS was contextual and authentic. Within the laboratory, scientists assessed the preservice teachers' understanding of the science processes and content during the laboratory experiences. Each squad of preservice teachers, along with their scientist, conducted experiments in the laboratory. The small group interactions allowed the scientist to evaluate student understanding during each experiment and guide and direct their activities according to their level of skill and understanding. In this sense, the assessment was informal and a natural part of the laboratory environment. For example, in the materials science laboratory, the preservice teachers used a glove box to prepare samples. The scientist directly observed their work. One preservice teacher had difficulty functioning in the glove box and preparing the sample. The scientist first tried talking the preservice teacher through the difficulty. When this did not work, the scientist had the preservice teacher move away from the glove box and observe the scientist prepare the sample. However, the scientist did not fully complete the task and had the preservice teacher return to the glove box to finish preparing the sample. This type of just-in-time instruction directly met the science learning needs of the preservice teachers. In contrast to most of their prior science experiences, the ongoing assessment of their skill and understanding was non-threatening yet personally meaningful for the preservice teachers as they worked to makes sense of the science content.

In the elementary science classroom, the classroom teacher informally assessed the preservice teachers' knowledge of and skills in teaching. The primary means of assessment were the guided reflection sessions that occurred after the preservice teachers' observed the science classroom. Through discussion, classroom and preservice teachers were able to decompose the instructional events that had occurred during the science lesson. In these sessions, the classroom teachers were able to make their thinking visible to the preservice teachers. For example, after a lesson on gravity, the classroom teacher asked the preservice teachers to describe the types of questions she asked the students during the lesson. The preservice teachers noted that her questions asked them to explain and justify their reasoning (for a related issue see the chapter by Roth, p. 20). The classroom teacher then pointed out that she seldom provided specific answers to students working on science experiments. Instead, she responded to student questions with guiding questions so they could discover answers on their own. The preservice teachers were able to first observe this behavior on the part of the teacher and then hear the teacher describe why he/she did this. The guided reflection sessions provided opportunities for the preservice teachers to ask questions and increased the level of accountability the preservice teachers had for analyzing the instructional events.

As the preservice teachers' knowledge of teaching science progressed, so did the nature of their involvement in the classroom. The classroom teachers incrementally increased the preservice teachers' instructional roles according to their skill and knowledge. Initially, preservice teachers worked with small groups of students on well-defined tasks and progressed to formally implementing parts of a science lesson. For example, early in the TEAMS project, the preservice teachers would work with one group of students who were performing a science experiment designed by the classroom teacher. The preservice teacher would assist the students in setting up the experiment and recording the results. Again, the guided-reflection sessions made assessment of these laboratory experiences a natural part of teaching and were a non-threatening means of developing the preservice teachers.

The culminating events for assessing the preservice teachers' growth were the science lessons or mini-units they developed and implemented. The content of the lesson was based upon the preservice teachers' science laboratory experiences. In addition, the lesson had to be integrated into the learning activities of the elementary class. During the first week of TEAMS, the preservice teachers were informed that they needed to develop science lessons based on their laboratory experiences and implement all or part of the lessons in the classroom. Because the laboratory and classroom experiences were unique for each squad, general guidelines for the lesson plans were provided. The final lessons plans were to include: statements concerning the science background and the connection between laboratory and classroom, the lesson objectives, materials needed, concepts transferred from laboratory, the lesson procedures, and the lesson evaluation. The specific roles of each squad member are outlined in Table 1.

Table 1: Participant Roles in TEAMS Final-Lesson Development and Implementation

Preservice Teachers	*Classroom Teachers*	*Teacher Educators*	*Scientists*
Brainstorm lesson topics	Brainstorm lesson topics	Brainstorm lesson topics	Brainstorm lesson topics
Define specific lesson topics	Provide curriculum resources (e.g. school curriculum guides, list of upcoming units)	Provide curriculum resources (e.g. National Science Education Standards)	Provide rich science laboratory experience s that can be transferred to classroom
Identify how science concepts fit into curriculum	Assist in defining age-appropriate activities	Assist in defining age-appropriate activities	Help preservice teachers understand broad science concepts involved in laboratory research
Design lessons and activities that lead to meaningful science learning	Assist in refinement of lesson activities	Assist in refinement of lesson activities	Provide technical guidance for lesson (e.g. samples, equipment)
Produce final lesson plans	Review pedagogical component of lesson plan	Review pedagogical and science components of lesson plan	Provide technical guidance for lesson (e.g. samples, equipment)
Implement lessons	Provide pedagogical assistance during lesson implementation	Assist in relating laboratory science to school science activities	Provide technical assistance during lesson implementation
Critically review lesson implementation	Provide feedback on lesson implementation	Provide feedback on lesson implementation	Provide feedback on science component of lesson

Although the students produced lesson plans, the plans were not the primary focus of the assessment. The preservice teachers' implementation of the lesson was the focus of the assessment. Consistent with the assessment procedures used in the classroom, guided-reflection sessions (involving the squad scientist and teacher educator) were used to assess the lesson implementation.

EVALUATION OF TEAMS

The focus of TEAMS was not simply to provide interesting science laboratory and classroom experiences for the preservice teachers. Rather, a central purpose of the program was to facilitate an experience where the preservice teachers would see a connection and possible avenue for transferring the science and their laboratory experiences to the classroom.

Evaluation Procedures

To better understand the TEAMS model and how their experiences impacted the preservice teachers' development, case studies of individual squads were conducted. The case study methodology was used because it facilitates the in-depth investigation of instances of a phenomenon (e.g. preservice teachers' development as science educators) in its natural context from the perspective of the participants involved in the phenomenon (Gall, Borg, & Gall, 1996). Data gathering techniques used in the case studies included individual and focus group interviews, site visits, surveys, and e-mail journals. Preservice teachers participated in individual and group interviews and journaling activities, and they were observed as they worked in the science laboratories and the classrooms by the project evaluator. Table 2 provides a detailed summary of the evaluation plan for the project.

The individual interviews with the preservice teachers were conducted from an interview schedule that included sixteen specific questions. These interview questions were divided into four categories: student description of the project, student evaluation of what they had gained from the project, student expectations for science teaching in the future, and general student reactions.

Focus groups were conducted separately for the preservice teachers, scientists, teacher educators, and the classroom teachers. The purpose of each of the focus groups was to encourage each of the participants to reflect on the meaning of the experience. Focus group questions prompted the participants to discuss their experiences in TEAMS and provide suggestions for future projects.

The 48-item science attitude survey administered to the preservice teachers consisted of four subscales: usefulness of science, confidence in teaching science, effectance motivation in science, and perceptions of science and scientists. The science attitude survey used a five point Likert scale (5 = strongly agree, 4 = agree, 3 = undecided, 2 = disagree, 1 = strongly disagree). The cumulative total of the items on each attitude subscale was calculated for each preservice teacher. These data were used to determine the preservice teachers' attitudes toward science.

The project evaluator, a Ph.D. graduate student with a specialization in evaluation, administered the science attitude survey and conducted the individual interviews, focus groups, and site visits. She collected the data from each group and analyzed those data for major themes. These major themes are summarized below.

Table 2: Evaluation Plan for TEAMS Model

Evaluation Goals	*Measurement Methods and Instruments*	*Subjects*	*Administration of Measurements*
Describe TEAMS model and implementation	site visits to laboratories and classrooms	preservice teachers	several times throughout program
		scientists classroom teachers	during site visits (informal)
	interviews	teacher educators	
			end of program (formal)
Describe preservice teachers' understanding of nature of science and the nature of effective elementary science instruction	email journals	preservice teachers	throughout program
Describe connections preservice teachers make between laboratory and classroom	email journals	preservice teachers	throughout program
	focus groups	preservice teachers scientists classroom teachers teacher educators	end of program
Determine impact of model on preservice teachers' confidence in ability to teach science	science attitude survey	preservice teachers	pre-post program
	email journals	preservice teachers	throughout program

Results

The TEAMS experience expanded the preservice teachers' conceptions of science. This was a major theme that emerged from comments during focus groups and interviews. The science laboratory component authenticated and legitimized science for the preservice teachers, most of whom, prior to TEAMS, acquired knowledge of, and interactions with, science through lectures, plug-and-chug formula equations, and textbook-driven experiments. ("We see a different part of science than we are used to in the classroom; we get to see what the scientist does, how they go about what they do, what questions they are trying to answer".)

The elements of the science laboratory experience most cited by the preservice teachers in both individual interviews and focus groups were the setting and the people. The laboratory setting allowed the preservice teachers to be immersed in an environment where the on-going practice of science was the norm; they were exposed to a culture where the practice of safe, rigorous, objective scientific inquiry is the highest value. In this environment, the preservice teachers used the research equipment, engaged in meaningful science processes, and interacted with and observed the various participants. Having the opportunity for regular small group and one-on-one interactions with a scientist helped make the science laboratory experience meaningful for the preservice teachers. The scientists created the atmosphere for the laboratory experience. The preservice teachers characterized the atmosphere as practical, personal, and profitable.

Practical: The activities of the laboratory required the preservice teachers to take action; they measured elements, made samples, mixed compounds, etc. The activities were conducted as a group, thus enabling and encouraging conversations. Frequently, one question lead to another, which often led to still another. This allowed the preservice teachers to engage in science, verbalize their actions, and discuss the implications of the phenomenon with an expert. ("We can ask her questions, she is up on such a high level, but no question is too stupid. We get some really interesting discussions going".)

Personal: Because each squad consisted of three or four preservice teachers, the science experience became more personal to them. Working with the scientist on an activity for which each member had ownership, the squads developed a bond and a sense of camaraderie. ("Maybe this sounds corny, but we were a team, reaching out, working together".) In addition, the preservice teachers had the opportunity to know the scientist on a more personal level. This helped them view scientists as regular people with whom they have something in common. Although the number of students completing the Science Attitude Survey is currently too small to apply inferential statistics, preliminary survey results indicate that student perceptions of scientists became more positive during the course of the project.

Profitable: The preservice teachers considered the time in the laboratory to be productive; not so much in the sense of developing a product, but in the sense of fostering attitudes toward learning. ("I saw the value of making mistakes. We started out by making mistakes... lots of them. We learned a lot. Everyone learns that way".) To assist the preservice teachers in making connections between the laboratory science and teaching science, each squad was to design and implement a lesson based on the science concepts they learned in the laboratory. It was evident through the lessons developed by the preservice teachers that they identified connections between the laboratory and the classroom. For example, in the laboratory, one squad had investigated the electrical properties of various materials and had observed lessons on electricity in the classroom. To integrate the experiences, they designed and taught lessons using the concepts of conductivity to explore the electrical properties of common metals with fifth grade students. In addition to the lessons, the comments of the preservice teachers indicated that they saw similarities between science and education. ("I really enjoyed Barb.... She is a teacher even though she doesn't realize it.... We can ask her all kinds of questions",) The preservice teachers also referred to the practicing teachers as scientists. ("Suzanne is a scientist. She is always testing out hypotheses 'I wonder if...? Did you see what I did differently in the first class? Why do you think I did?'")

The practicing classroom teachers and the scientists influenced the science and education philosophies of the preservice teachers. ("When I look back through it, and I'm interacting with my kids, I can see myself doing things that she has done.

She's been a great model for me and what I want to do with my students" or "When I was taking science, I don't remember a lot of people liking science. Now I've had this experience of actually working with scientists, seeing what they do.... I'll be able to tell my students 'this is what I did when I was in the lab.'") Again, the science attitude survey data reinforce these results. The preservice teachers showed positive gains in items measuring the usefulness of science and their confidence in teaching science. Survey results also suggested that by the end of TEAMS, preservice teachers had much higher expectations for the type of science activities that they could lead at the elementary level.

TRANSFERABILITY OF THE TEAMS MODEL

One of the purposes of this chapter is to describe the TEAMS model and project so that other teacher education institutions can implement similar projects. Clearly, the fact that the Ames Laboratory is located within one block of the College of Education at Iowa State University helps facilitate the implementation of this project. Although other schools or colleges of education will probably not have scientific resources quite so close at hand, the model can be extended to different types of collaborations. Most colleges or schools of education have science research facilities in close proximity. These research facilities might be within the university, or they might be part of private industry. As more and more businesses are working to collaborate with and support education (especially in the areas of mathematics and science), a project similar to TEAMS may be a useful way to begin relationships and establish linkages.

Scientists in our project spent approximately three hours per week on the project, including preparation time and time with the preservice teachers, and were not paid for their time. Scientists' evaluations of the experience indicated that the project was a valuable learning experience for them, and almost all of the scientists who have participated in the project have indicated enthusiasm for future participation in TEAMS. We believe that others implementing a TEAMS-like project will find similar attitudes and collaboration from local scientists.

Classroom teachers in the project were paid small stipends for their participation ($500 per semester) and also provided budgets for project equipment ($1,000 per semester). Teacher participants reported that the experience working with the scientist was extremely useful for them and many TEAMS teachers and scientists have continued their collaborations far beyond the bounds of the project. In one case, the classroom teacher now visits the scientist's college-level classes to provide feedback on his teaching techniques. In another case, the participating scientist now provides regular laboratory experiences for the classroom teacher's fifth grade class and has become something of a resident scientist in her class each semester.

In a typical semester, we have 16–20 preservice teachers in the TEAMS. Although it would be difficult to scale up a project like this to involve all teacher education students in an institution, the experiences from TEAMS may help create teacher-leaders in the area of science. According to the science attitude survey, the preservice teachers reported increased confidence and enthusiasm for teaching science and believed they would be capable teachers of science in their first year. Preservice teachers in the program share their experiences in TEAMS with peers, and thus, additional preservice teachers receive some benefits from the program.

Many aspects of the TEAMS model can be adapted to other teacher education programs without extensive external funding. Although external funding clearly helped establish the project, it now operates without significant stipends or support for the scientists or the classroom teachers. In fact, TEAMS has helped create a collaborative base among scientists, classroom teachers, teacher educators, and preservice teachers that has served to facilitate additional projects. For example, numerous successful NSF grant proposals have grown out of relationships established in TEAMS.

CONCLUSION

Grounded in cognitive apprenticeship theory and principles of science education reform, the TEAMS project provides a useful model for teacher education institutions seeking to provide high quality science content and pedagogy experiences for their preservice teachers. In the TEAMS model, preservice teachers have authentic experiences with expert scientists and expert classroom teachers and have opportunities to reflect upon and apply these experiences. As the project evolved both scientists and classroom teachers were asked to model expert behaviors for the preservice teachers to observe and then to provide scaffolding to help the preservice teachers interpret and apply these experiences. Since the science experience of the scientists was markedly different than that of the preservice teachers, the challenge of providing scaffolding to make the modeling experience meaningful was large. With the teachers, the modeling of expert behaviors was easier for the preservice teachers to interpret and understand. Thus, the scaffolding portion of the process was not as complex as it was for the scientists. In the laboratory and the classroom settings, some fading was used as preservice teachers assumed increasing responsibility and independence and as their knowledge and confidence grew. Scientists and teachers acted as effective coaches for the preservice teachers as they designed and implemented lessons based on their laboratory experiences.

Results from TEAMS indicate that the cognitive apprenticeship model provides a useful framework for designing, implementing and revising a science teacher preparation program involving scientists, master science teachers, and teacher educators in a reflective collaboration. Data from the project evaluations suggest that preservice teachers who have participated in TEAMS have achieved expanded knowledge of science, increased confidence in their ability to teach science, increased interest in teaching science, and the potential to develop as leaders in science education. Master teachers participating in the project report similar outcomes, while scientists cite increased knowledge of K–6 education and science pedagogy as project outcomes. The TEAMS model as a whole or in portions provides generalizable results for other teacher education programs striving to incorporate more authentic science learning experiences for preservice teachers.

REFERENCES

American Association for the Advancement of Science. (1990). *Science for all Americans* (Project 2061). New York: Oxford University Press.

Anderson, R. D., & Mitchener, C. P. (1994) Research on science teacher Education. In D. L. Gabel (Ed.), *Handbook of research on science teaching and learning* (pp. 3-44). New York: Macmillan.

Brown, J. S., Collins, A., & Duguid, P. (1989) Situated cognition and the culture of learning. *Educational Researcher, 18*(1), 32-42.

Collins, A., Brown, J. S., & Holum, A. (1991). Cognitive apprenticeship: Making thinking visible. *American Educator, 6-11*, 38-46.

Collins, A., Brown, J. S., & Newman, S. E. (1989). Cognitive apprenticeship: Teaching the crafts of reading, writing, and mathematics. In L. B. Resnick (Ed.), *Knowing, learning, and instruction: Essays in honor of Robert Glaser* (pp. 453-494). Hillsdale, NJ: Lawrence Erlbaum Associates.

Doyle, W. (1990). Classroom knowledge as a foundation for teaching. *Teachers College Record, 91*, 347-360.

Gall, M. D., Borg, W. R., & Gall, J. P. (1996). *Educational research: An introduction* (6th ed.). White Plains, NY: Longman Publishers USA

Goodlad, J. I. (1990). Studying the education of educators: From conception to findings. *Phi Delta Kappan, 71*(9), 698-701.

Greeno, J. G., Moore, J. L., & Smith, D. R. (1993). Transfer of situated learning. In D. K. Detterman & R. J. Sternberg (Eds.), *Transfer on trial: Intelligence, cognition and instruction* (pp. 99-167). Norwood, NJ: Ablex.

Grossman, P. L, Wilson, S. M., & Shulman, L. S. (1989). Teachers of substance: Subject matter knowledge for teaching. In M. Reynolds (Ed.), *Knowledge base for beginning teachers* (pp. 23-36). New York: Pergamon.

National Research Council. (1994). *Standards for the professional development of teachers of science.* Washington, DC: National Academy Press.

National Research Council. (1996). *National science education standards.* Washington, DC: National Academy Press.

National Science Teachers Association. (1993). *Revitalizing teacher preparation in science—An agenda for action.* Washington, DC: Author.

Radford, D. L. (1998). Transferring theory into practice: A model for professional development for science education reform. *Journal of Research in Science Teaching, 35*, 73-88.

Raizen, S. A., & Michelsohn, A. M. (1994). *The future of science in elementary schools.* San Francisco: Jossey-Bass.

Roth, W.-M. (1996). Teacher questioning in an open-inquiry learning environment: Interactions of context, content, and student responses. *Journal of Research in Science Teaching, 33*, 709-736.

Roth, W.-M. (1998a). Science teaching as knowledgeability: A case study of knowing and learning during coteaching. *Science Education, 82*, 357-377.

Roth, W.-M. (1998b). Teaching and learning as everyday activity. In B. J. Fraser & K. G. Tobin & (eds.), *International handbook of science education* (pp. 169-181). Dordrecht, Netherlands: Kluwer Academic Publishing.

Trumbull, D. (1990). Evolving conceptions of teaching: Reflections of one teacher. *Curriculum Inquiry, 20*. 161-182.

A PROBLEM-BASED LEARNING APPROACH TO SCIENCE TEACHER PREPARATION

Raymond F. Peterson[1] & David F. Treagust[2]
1University of Adelaide, 2Curtin University of Technology

INTRODUCTION

Beginning in the medical profession, problem-based learning (PBL) has now been used in a range of university programs (Boud & Feletti, 1991), including nursing, design, engineering, architecture, science, pharmacy, social work and psychology (Ryan & Little, 1996). The impetus for change to a PBL curriculum was often based on the need to address the balance between the academic disciplines and the professional knowledge required for individuals to be a member of a profession (Boud, 1985). As a result, programs shifted from subject-oriented studies to a more integrated, contextual, and holistic understanding of knowledge and professional practice. This shift in emphasis of learning in the field also assumed that learning a large body of knowledge is not sufficient to be an expert in the field (Margetson, 1993; Roth, this volume). Although PBL has now been used extensively in a range of higher education, very few studies report on the use of PBL in a teacher education context (Chappell & Hager, 1995).

Problem-based learning addressed a number of issues in relation to learning and teaching in higher education. For example, teachers concerned with the transmission of knowledge were often convinced that PBL was a better alternative for teaching and learning. It took into account how students learn, developed students' skills to learn quickly, effectively, and independently rather than acquiring a body of knowledge at graduation, and also had high credibility with many professionals about what happens in their field of activity (Boud & Feletti, 1991; p.17). In addition, with the ever-increasing knowledge base in all professions, it was recognized that for many students irrespective of the discipline, the knowledge acquired in their higher education program was likely to change over their professional careers. For this reason, students needed to have acquired the ability and appreciate the importance of being lifelong and self-directed learners to adjust to possible changes in their profession (Engel, 1991). PBL was identified as a possible approach to address these concerns, as it was identified to develop students' knowledge bases in their profession, and their reasoning and problem-solving abilities associated with the discipline. PBL provided opportunities for students to be more self-directed in their learning, and increased students' motivation to learn when compared to a more traditional teaching situation. The move to a PBL approach in teacher education is supported by views expressed in the *National Science Education Standards* (NAS, 1995). The *Standards*

D.R. Lavoie and W.-M. Roth (eds.), Models of Science Teacher Preparation, 49–66.

suggest that teachers need the skills and ability to be lifelong learners, reflect on their own practice, and take responsibility for their own professional development.

Although PBL has not been extensively used in science teacher education, the integration of knowledge (in this case knowledge of content, curriculum and learners) with the pedagogy of teaching, and the need to have self-directed and lifelong learners in the teaching profession places preservice science teacher education in the position of being a suitable context for a PBL program. Problem-based learning also addresses the concern expressed in the *National Science Education Standards* that teacher preparation courses focus on technical skills rather than decision-making, theory and reasoning. To overcome this situation, the *Standards* reported that prospective teachers need to be engaged "in active learning that builds their knowledge, understanding, and ability" (NAS, Chapter 4, p. 2). In this chapter, we outline a framework for the use of a PBL approach in a science teacher education context and describe its implementation in a science teacher preparation program.

Background

Typically, PBL programs are based around carefully selected, designed and written cases to meet various objectives of a curriculum. The cases contain trigger material to encourage students to begin exploring particular areas as they attempt to understand the issues or problems identified in the case scenario. The most common format for cases is a paper description of the problem, although problem scenarios have been presented as videotapes, audiotapes, role-plays, or computer simulations (Lovie-Kitchin, 1991). Cases can vary in duration from up to four tutorial sessions, through to a case extending over a semester. Professional education programs, such as science teacher education, are well suited to a PBL approach as there is a wealth of case scenarios available from within the practice of the profession, as well as from the research literature related to the practice of science education in school settings.

Students encounter the problem first in the learning sequence before any planned preparation or study in the area has occurred. This is a significant shift in the learning process, and is important as students determine what they need to learn. In presenting a case to students, there are five common elements that characterize a PBL environment (Barrows & Tamblyn, 1980):

- The problem is presented to students in the same manner as it would occur in a professional context. This provides further relevance to the students to engage in the case.
- The students work with the problem in a manner that enables them to use their ability to reason, apply knowledge, and to challenge and evaluate their understanding appropriate to their level of learning.
- Areas where learning is needed to address the problem are identified in the process of working with the problem and these areas are used to guide individual study to be completed between sessions.
- The skills and knowledge acquired in this study are applied to the problem to evaluate the effectiveness of the learning and to reinforce learning.
- The learning from working with the problem in a group and through individualized study is summarized and integrated into the student's existing knowledge and skill base.

There is considerable scope for members of a course design team to adapt PBL to their context provided that they recognize that the learning is much more student-centered, the learners are active participants in learning, and the learners' views about what should be learned are considered. If these aspects are considered then the learning should be more meaningful to students and not just result in the development of a body of unrelated knowledge (Boud, 1985; Boud & Feletti, 1991; Engel, 1991). For these reasons, early applications of PBL placed significant emphasis on the small group setting as the environment to enhance learning (Barrows & Tamblyn, 1980). However, more recent PBL programs have used a combination of large and small group learning environments to accommodate timetable and course constraints in an existing curriculum (Cleary 1996; Heycox & Bolzan, 1991; Woods, 1991).

Framework

To use PBL in a science teacher education context, a framework was developed that described (a) the knowledge base for teaching and (b) the pedagogical reasoning associated with the application of the knowledge to the profession. For the discussion in this chapter, the knowledge base for teaching will only focus on three components:

1. *Science Content Knowledge*. This was defined as the factual information, organizing principles and central concepts of a discipline. (Grossman, Wilson & Shulman, 1989)
2. *Knowledge of Curriculum* focussed on developing knowledge of curricular alternatives for a topic, and the curriculum materials in the topic area (Shulman, 1986, p. 10). These materials included textbooks, instructional materials, audio, video and computer software. (Marsh, 1992)
3. *Knowledge of Learners*. This focussed on student learning in science, and the development of strategies to explore learners' prior knowledge before teaching a topic at the elementary level, and to view science learning as a process of knowledge construction starting from this existing knowledge base. (Treagust, Duit, & Fraser, 1996)

These three knowledge-base components are only part of overall descriptions of the knowledge base for teaching often described to include subject matter knowledge, general pedagogical knowledge, curriculum knowledge, pedagogical content knowledge, knowledge of learners, knowledge of educational contexts, and knowledge of educational ends, purposes and values (Reynolds, 1992b; Shulman, 1987; Shulman & Sykes, 1986; Wilson, Shulman, & Richert, 1987).

As part of working on a PBL case, future teachers develop reasoning and problem-solving skills appropriate for their profession. Therefore, in addition to developing a knowledge base for teaching, preservice teachers develop the pedagogical reasoning ability to use this knowledge for making decisions about teaching and learning in a classroom situation. In recent years, various descriptions of pedagogical reasoning have been described in the research literature. For example, Reynolds (1992a) described this pedagogical reasoning process as a teaching task framework which comprised four domains; a pre-active domain (e.g., understanding of the topic), an inter-active domain (e.g., implementing and adjusting plans during teaching) and post-active teaching tasks (e.g., reflecting on own and student performance). A fourth, administrative domain is based on the tasks required of teachers in the school setting.

In another model (Wilson et al., 1987) six components were described in the reasoning process, namely:

1. *Comprehension*: Teacher understanding of the ideas to be taught and the educational purposes of the topic/subject.
2. *Transformation*: Comprehended ideas are transformed by the teacher for use in a particular classroom setting. This includes critical interpretation of text materials, identifying ways of representing ideas, selecting appropriate teaching methods, adapting and tailoring ideas to the particular class group.
3. *Instruction*: The act of teaching. This includes organizing and managing the class and students, presenting clear explanations, interacting with students, questioning and evaluating.
4. *Evaluation:* This includes both the evaluation of student learning and the teacher's own teaching performance, materials employed, etc.
5. *Reflection*: The review of the events and accomplishments that occurred during the lesson.
6. *New Comprehension*: New understanding of subjects, learners, purposes and pedagogy through the process of teaching.

This proposed model of pedagogical reasoning—although it implies a sequence—did not assume that all teachers would follow the fixed stages as described. It is likely that teachers will interchange between each, omit a component, or accomplish two components simultaneously (Reynolds, 1992a; Shulman, 1987).

In this chapter, we focus on the pedagogical reasoning model proposed by Shulman (1987) and Wilson et al. (1987). The structure articulated with the plans for the proposed PBL science teacher education program. The evaluation, reflection, and new comprehension stages are also in keeping with the expectations of the *National Science Education Standards* that "teachers of science develop the skills to analyze their learning needs and style through self reflection and active solicitation of feedback from others" (NAS, 1995, Chapter 4, p. 9).

As reported in the *National Science Education Standards* (NAS, 1995, Chapter 4, p. 6) "skilled teachers of science have special understanding and abilities that integrate their knowledge of science content, curriculum, learning and students which results in the pedagogical content knowledge they use in the teaching process". The *Science Standards* further add that in learning to teach science teachers needed opportunities to analyze components of pedagogical content knowledge, namely science, learning, and pedagogy. Other studies also recommended to give pedagogical reasoning a higher priority in teacher education programs (Kennedy, 1990; McDiarmid, Ball, & Anderson, 1989; Reynolds, 1992b; Shulman, 1987). One strategy for achieving this was for preservice teachers to "experience... pedagogy first as learners.... This experience challenges their pre-existing scheme for teaching and learning. The subsequent cognitive conflict allows the accommodation to the new pedagogical conception" (Stofflett & Stoddart, 1993, p. 45). A problem-based approach to preservice teacher education addresses these issues through the integration of knowledge with reasoning in relevant professional and educational contexts.

PROBLEM-BASED LEARNING IN SCIENCE TEACHER EDUCATION

Background

The shift to a more problem-based science education program was based on identified weaknesses in the existing program, and weaknesses reported more broadly on science education. These included the poor science knowledge base of elementary science teachers (Kruger & Summers, 1988; Speedy, 1989) and weaknesses in their knowledge base for teaching and pedagogical reasoning ability. For example, we earlier reported that first-year preservice teachers focussed in their pedagogical reasoning only on the comprehension of content knowledge and the transformation stage (Peterson & Treagust, 1992). Third-year preservice teachers demonstrated more of the six pedagogical reasoning stages but individuals were identified as having specific areas of weakness. The pedagogical reasoning ability of the first-year group was particularly important in the development of a PBL science education program.

The PBL Science Education Program was part of a compulsory second-year two-semester science education unit in a three-year bachelor of teaching (elementary) degree. In the existing two-semester science education unit, preservice teachers participated in a weekly two-hour session that was often a combination of a workshop, mini lecture, tutorial, and science-based activity. They also completed two school placements, for three and four weeks respectively. In the first-year, all preservice teachers had completed a one-semester science education subject and spent a total of four weeks in a school setting. During this time they made general observations on classroom teaching and organization and taught (at least) one science lesson to a small group of elementary students.

Design of the PBL Program

The program centered on the three knowledge base components, namely science content knowledge, science curriculum knowledge and knowledge of learners, and the six stages of the Shulman's pedagogical reasoning model. In designing cases, the school-based experiences with students in classrooms (combined with the extensive literature on student learning in science) offered valuable sources of information for the development of case scenarios. These cases were to be completed within a university setting. Other cases could be based on the classroom setting in which the students completed their school placements. Here is a typical university-based case scenario:

> *Background:* You are teaching in an upper elementary grade and have just completed some exploration work on seeds and flowers. In the exploratory activities students looked at some different flowers. In one activity students had four flowers from the same plant but at different stages of development and were asked to put them in a sequence. This led to a number of questions from the class. The questions included the following: What are the parts of the flower? Why do flowers have pollen in them? Why are flowers different colors? Why do some plants have flowers and some don't? What

are seeds made of? Do all plants have seeds? Can the seeds grow into plants? and Can you grow things without seeds?

Your problem: As the teacher you intend to use this information to plan a series of investigations for your class. It is likely that science activities will need to be tried as part of this preparation phase. After the preparation phase each person will be teaching his/her topic to a group of his/her colleagues. This is for three reasons: To enable your colleagues to develop an understanding of the topic; to enable your colleagues to see a possible teaching approach for the topic; and to enable you to trial some of your ideas and receive some feedback on the teaching approaches you have selected. It is important to consider alternative teaching approaches and not simply "hands-on" activities to explain your ideas.

The scenario meets a range of purposes. First, students are to develop an appreciation of the importance of beginning with elementary students' prior knowledge of the topic. Beginning second-year preservice teachers tended to ignore the prior knowledge of students in the teaching process, as they had mostly taught single lessons in the school setting. Consequently, the main emphasis in these single sessions was on the teaching of a concept with little (if any) recognition of the students' prior knowledge. Second, students were to develop an understanding of the science concepts relevant to the topic. The content knowledge topics for these university-based problems were selected by the researchers and were based on science topics studied as part of the second-year science education program. The main shift in emphasis was that the preservice teachers would be required to carry out their own research to develop their understanding of the science content. Third, students are to identify and evaluate suitable curriculum resources. Students had little experience at identifying, reviewing, and evaluating curriculum resources, and linking this to teaching. Fourth, students were to plan a suitable teaching approach for the topic for the classroom situation. Depending on their prior experiences, participants varied from being very teacher-directed through to student-directed in their teaching. They were expected to explore possible teaching approaches that would be suitable for their topic, and elementary student age and ability. Fifth, students were to use all six stages of the pedagogical reasoning process. This was an early case in the science education program and an important aspect of the planning was for students to develop their pedagogical reasoning as students did have limited pedagogical reasoning ability (Peterson & Treagust, 1995). A peer teaching component was included as part of this case, as students did not have access to elementary students to trial ideas and strategies for teaching. This peer teaching was important because it allowed each person to then evaluate, reflect and develop new comprehension of the topic based on this teaching experience, and by doing this they had the opportunity to use all six stages of Shulman's (1987) pedagogical reasoning.

An outline of some possible PBL case scenarios is given in Table 1. For each case, the science content and associated curriculum information, and grade level to be the focus of the case can be decided by the staff teaching in the program. The level of complexity of the case would depend on the prior knowledge and expertise of the elementary students.

Table 1. Outline of some possible PBL case scenarios

	Content Knowledge Base	*Pedagogical Knowledge Base*
1	Science content knowledge; Pedagogical reasoning process; Introduction to PBL approach; Working in small groups; Strategies for exploring prior knowledge in science; Recognizing that students have different understanding of the same science topic; Self-directed learning;	Students (in a small PBL group) explore the understanding of group members (approximately 6 in another group) for the science topic. Based on this information, the PBL group develops activities to improve the understanding of their colleagues. Activities are presented to colleagues. After completion, the PBL group considers the evaluation, reflection and new comprehension stages of pedagogical reasoning.
2	Science content knowledge; Pedagogical reasoning; Student learning in science; Building science activities on students' existence knowledge base and questions; Knowledge of curriculum associated with the science topic; Evaluating teaching strategies for teaching the science topic a particular grade level; Self-directed learning.	Using the given prior knowledge of elementary students for a science topic, the PBL group develops the science content and curriculum knowledge for the topic. Approaches for teaching the topic at a particular grade level are identified. Reporting to peers provides opportunities for evaluation, reflection and new comprehension
3	As above, but would also include gender-inclusive teaching approaches as part of the case (especially if the science topic focussed on, for example, machines). In this case, the PBL group would spend at least two sessions with elementary students—the first to identify prior knowledge, the second to implement planned activities.	The PBL group meets with a small group of elementary students to identify their understanding in a science topic. Appropriate learning activities with the group are then planned, implemented and evaluated
4	As described in case 2, but with the additional emphasis of students' understanding the inter-relationship between science, technology, mathematics etc.	The PBL group integrates science with one or more subject areas based on the prior knowledge and expertise of an elementary science class.
5	Science content knowledge; Pedagogical reasoning; Student learning in science; Building science activities on students' existence knowledge base and questions; Knowledge of curriculum associated with the science topic; Evaluating teaching strategies for teaching the science topic a particular grade level; Self-directed learning	Completed as an individual activity. The PBL problem in this case is a real classroom teaching situation in which the pre-service teachers are teaching science.

Implementing Cases with a Student Group

Small-group student-centered learning is an integral component of many PBL programs. Tutors have a significant role in facilitating the tutorial, and in teacher education this would include guiding the pedagogical reasoning process. In addition, the small-group process encourages collaborative learning, which in a teacher education context is developing the skills and attitudes in participants to take responsibility for their ongoing professional development (NAS, 1995). However, for this PBL program, the students worked in small groups without a tutor. Additional, paper-based questions were developed to enable students to facilitate their own small-group process. These guiding questions, which correspond to the six pedagogical reasoning stages, were as follows:

> Here are some questions your group should consider. Your group may also have other questions that need to be investigated. The questions are only listed to guide you.
>
> 1. *Comprehension:* What is your knowledge in this topic area? What do you need to find out? How, when, and where will you find out this information? Do the science ideas link together? What is the purpose of your topic for students?
> 2. *Transformation:* What curriculum materials are available on the topic? How will you use these materials? What curriculum materials will be used? How will the work be sequenced? How will you explain some of the ideas to students? Will you use demonstrations, examples, analogies, etc.? What teaching approaches best suit this topic? How will you cater for student interests/abilities?
> 3. *Instruction:* How will classroom teaching be organized? What questions will you ask during teaching? What teaching strategies will be used during the lesson? How will the students in the class be organized? How will you cater for early finishers?
> 4. *Evaluation:* How will you check for student understanding? What will you evaluate in your teaching performance? How will you evaluate your teaching performance?
>
> And after you have taught this...?
>
> 5. *Reflection:* How effective was the teaching for you/the group? What aspects went well for you /the group? What suggestions were made by your peers? What changes will you make and why? What are my strengths? What are my weaknesses?
> 6. *New Comprehension*: What new understanding of science have you developed? What new understanding of planning have you developed? What new understanding of teaching this topic have you developed? What did you learn from evaluating the person's understanding?

Students were not expected to complete these questions. However, they could use them as a guide if they were unsure where to proceed on the case. Small groups of four to six students worked on each of the cases. The teacher involved in the program acted as a facilitator for all of the groups. Thus, each group had to work independently for some time. We assumed that the participants in this study had sufficient skills and training to work in small groups, based on their experiences and understanding of the theory and practice of group dynamics from other subjects in the bachelor-of-teaching degree. Developing group members' skills to function in small groups is an important component of the problem-based–learning process. Some programs spend considerable time developing these skills in their student groups (cf. Woods, 1994).

Using a Journal to Promote Reflection-on Pedagogical Reasoning

To assist the preservice teachers in developing their pedagogical reasoning skills and knowledge base for teaching, guiding questions were also provided in a semi-structured journal. This strategy was particularly useful in the early stages of students using a PBL process and especially if the students need to work without the assistance of a tutor. Typical questions in the journal for this problem-based unit were as follows:

7. What questions from my problem did I consider this week? What did I learn? What still needs to be done? How will I do this? What are my strengths and weaknesses in this area?
8. What have I learned this week in relation to the science ideas in my topic? Do I understand the ideas? Can I see how the ideas are related? What do I still need to do? Can I see the purpose of my topic for students?
9. What ideas/thoughts/insights this week have helped my planning of science lessons? How useful are the curriculum materials? Can I understand them? Can I work out a suitable sequence for the topic? What alternative explanations have I developed? What else do I need to find out in planning?
10. What ideas/thoughts/insights this week have helped in developing ways of teaching this work to my group/elementary students? What questions might be useful? Why? What teaching strategies have I considered for this topic? Which ones will I use? How do I feel about my teaching at this stage?
11. What have I considered in relation to student evaluation? How will I know what has been learnt when I teach? What will I evaluate in my teaching of the group? Why?
12. Any other comments about my work?

Consistent with a PBL approach, which encourages students to be self-directed learners, these guiding questions were to promote reflective thinking. An added advantage of the journal is that it provides some insight into an individual's thinking, and this was particularly useful when tutors were not present in each group.

Assessment in a PBL Program

A PBL program is no different than any other program in that the intention of the assessment is to measure student performance against defined objectives, criteria or outcomes. However, PBL programs aim to integrate information from a variety of sources in attempting to solve the particular problem. Ideally, assessment approaches also should aim to reflect this view of integration.

There is considerable scope in the way in which a PBL program is assessed. For example, most PBL programs would aim to develop students' knowledge base and reasoning ability, and their ability to be self-directed learners. Journals, assignments, and the planning and evaluation of school-based teaching activities all provide opportunities to assess student performance against these objectives. The pedagogical reasoning criteria suggested in Table 1 could be used in the assessment process. Knowledge-base aspects in science or science education could be assessed using pencil-and-paper questions. However, test questions that model a problem-based approach are more likely to encourage students to focus more on reasoning rather than just rote memorization of facts.

The use of peer assessment and self-assessment against particular criteria could be used to assess students' ability to work collaboratively in a small group when working on the problem. Group and or individual presentations could be used as part of the assessment program. Grading differs in the PBL context: In some PBL programs, students are only given a pass/fail grade. The main argument for this approach is that a graded (A, B, C etc) environment encourages competition and distracts from the collaborative opportunities for students working in small groups.

AN EVALUATION OF A PBL SCIENCE TEACHER EDUCATION UNIT

This evaluation draws on data from one problem-based case where four different science themes, namely *Batteries and Bulbs*, *Eyes and Sight*, *Flowers and Seeds*, and *Spiders* were studied by different student groups (n = 4–6). The format used for the problem scenario for *Flowers and Seeds* was used for the scenarios for the other three themes. By using four different themes, the preservice teachers could teach part of their topic to a group of their colleagues who had not been part of the case, and to discuss their plans for teaching this unit to elementary students.

A case study methodology (Merriam, 1988) was used to collect data for the evaluation, and the methods included field observations, student journals, interviews with participants, questionnaires, videos of teaching and analysis of written documents such as lesson plans. The main purposes of the evaluation were to establish whether the PBL approach enabled preservice teachers in developing their pedagogical reasoning ability and the three components of the knowledge base for teaching (i.e. knowledge of content, curriculum, and learners).

Field observation data were collected by observing preservice teachers at work during the two-hour workshop sessions. All observations were recorded in a field journal. Journals were also kept by the preservice teachers during the problem-based units and used to identify preservice teachers' pedagogical reasoning and knowledge base development as the units progressed. Semi-structured interviews were conducted with a subset (n = 5) of the class group for the duration of this part of the PBL program. The preservice teachers were selected based on three criteria. These criteria included their ability to communicate effectively, representativeness with respect to gender, and representativeness with respect to grades. A review questionnaire was administered at the end of the problem-based unit.

DEVELOPMENT OF A KNOWLEDGE BASE FOR TEACHING

Science Content Knowledge

Prior to beginning the PBL program, preservice teachers often believed that science content knowledge was the most important aspect so that they had the confidence and ability to respond to student questions. ("The thing is, they're [the students] going to spring 'why is that going to happen?' on us. We'd look a bit silly if we just sit there" [David, Int P1-4, p. 6].)

In the PBL cases, the preservice teachers were able to develop their science content knowledge and build upon their existing knowledge in the topic that they investigated. They recognized when their science content was limited ("That's what really shocked us because we didn't know anything. I mean we knew there's a battery... there's a bulb but we just didn't have any idea of how they worked" [Mandy/Int 2.1/2]). Concept maps were used to assist the preservice teachers in identifying their science knowledge. Preservice teachers also recorded their understanding of the science content in the journals and teaching-unit summary. The level of reporting of science contents often reflected the detail that preservice teachers believed was necessary. For example, Karen provided detailed notes in her unit summary on the types of spiders:

> Hunters do not have webs. They catch their prey by running, jumping and pouncing. Hunters live in trees, the ground and houses. These spiders are generally brown or black. Spinners use their webs to capture their prey. They can be all sorts of colors and spin various types of webs. (Teaching Unit Summary/5).

Throughout the problem, preservice teachers were regularly reviewing and updating their science content knowledge, as they were able to identify and focus on specific aspects of their own learning needs. ("I have looked at a diagram of the structure of the eye and read and understood the functions of each part. I'm not too sure about long and short distance focussing so I need to find out about that" [Michelle/Journal-2/4].) Preservice teachers were able to identify what knowledge they required to understand as part of their work on the PBL cases. For each person, the level of knowledge varied and this in part was dependent on their prior knowledge in the topic area and the level of knowledge they believed was necessary to meet the requirements of the case. The participants valued group discussion while developing their science knowledge. ("Explaining information to others helps you to understand the concepts better. In groups it tends to be easier to stay on track, also if you don't understand how to experiment the other members of the group can assist you". [Jane, Journal-1/9].)

Elementary science practical activities were also important in developing an understanding of science content, as by trialling these activities preservice teachers were able to clarify concepts and to raise questions relating to their understanding of the ideas. ("The practical activities allowed me to actually see the concepts we discussed working. A visual presentation of the concepts related to air enable me to develop a mental picture much clearer than what I would have formed without seeing the activities, that is, I could understand [it] a lot better". [Jane, Journal-1/9].) It was evident that these preservice teachers were able to identify science concepts that needed further investigation. A comparison of pre- and post-unit concept maps demonstrated an increase in the number of concepts used and linkages between these concepts (e.g., Peterson & Treagust, 1995, 1998).

Knowledge of Curriculum

Initially, most preservice teachers only had limited knowledge of curriculum materials, with most of this knowledge acquired from specific science topics studied in their first-year science education subject or from the resources acquired through their teaching experience in the school setting. Therefore, one of the aims of this PBL program was to allow preservice teachers to explore a range of materials and evaluate the effectiveness of curriculum materials when deciding how they would teach a topic, either in the peer teaching situation or the elementary classroom.

The preservice teacher's often perceived improved understanding of science content helped in their understanding of the curriculum materials. ("Through my understanding of the [science] concepts it became easier to make links [with curriculum materials]" [Mandy/Journal-2/7]). They were also beginning to critically evaluate the effectiveness of some curriculum activities and resources. Sally, for example, suggested an alternative for an activity where two balanced and inflated balloons (one which is later burst) illustrate the idea that air has mass. Sally commented in this way:

> Air has mass. A difficult concept to deliver, because children can't feel the weight of air and because it is so minute [it] is difficult to measure. Rather than the activity we used I would suggest the following alternative. Attach two uninflated balloons to a metre rule. Suspend this so it is balanced. Inflate one balloon and notice that the balance is disturbed. This balloon is obviously heavier and the only thing that has been added is the air in the balloon Alternatively, if an electronic balance is available an uninflated balloon can be weighed, inflated, weighed again, and the results compared to prove that air does weigh something. (Sally/Journal-1/13)

The curriculum resources used by preservice teachers varied, and the range included books, practical activities, games, posters and audio-visual materials that would be applicable to their topic. In this PBL unit, all preservice teachers prepared a topic summary, which included the curriculum resources that were identified and evaluated as being useful for their particular topic. This self-directed curriculum exploration, with further opportunities for discussion in a group, enabled all preservice teachers to explore curriculum materials and to use them to develop their ideas and approaches for teaching the topic in the elementary setting.

Knowledge of Learners

At the beginning of second-year it was not uncommon for preservice teachers to have a teacher-directed approach to teaching and learning in elementary science. They made limited use of students' prior knowledge when planning a topic. This was in part due to the fact that they often taught only a single lesson in a discipline area during their school experience. Consequently, they had limited opportunities to develop science ideas over a period of time with a class group. Therefore, preservice teachers would decide what would be learned by the elementary students from their own perceptions of students' abilities ("I felt that yes, they're up to this stage, in a way it was a little vagueish ... but ... that's about their right level" [Linda, Interview P1-2, p. 7].)

The PBL program was designed to increase preservice teachers' awareness of the need to consider the learner's prior knowledge as part of the teaching/learning process. A review of their journals at the completion of the second problem scenario indicated that 15 of the 21 preservice teachers recognized the importance of having some understanding of the prior knowledge of their peers in deciding on a starting point for their teaching. The following comments typified their views on the value of assessing a student's prior knowledge of the science concepts:

> What the person already knew about the topic was very important information. It allowed us to view *what* [student emphasis] the person knew, how much they knew and it also allowed us to determine in which areas, if any the learners were misinformed and in which area we didn't have to go into great detail because they already had some prior knowledge. (Julie/Journal-1/15)
>
> Teaching this topic to children would have to begin with discovering how much they know or don't know about the topic. I may start with conferencing, for example, Who could tell me something about spiders? What spiders are poisonous? (Megan/Journal-2/5)

Overall there was a greater awareness of the need to understand the learners' prior knowledge of the topic when establishing a starting point for the teaching and learning in a given topic. The peer-teaching component of the problem-based units was instrumental in focusing preservice teachers on the need to consider the prior knowledge of the learners, irrespective of their age, when planning a teaching activity. Further evidence that the preservice teachers were beginning to incorporate the learners views into their science teaching comes from a four-week school-based teaching experience that followed this PBL unit. Fourteen of the 19 teachers participating in this experience used an exploratory science lesson to identify students' prior knowledge of the topic as the starting point for their science investigations with the class.

Development of Pedagogical Reasoning

At the beginning of second-year, preservice teachers' pedagogical reasoning ability was often limited to the comprehension and transformation stages, primarily as they had had few opportunities to teach science in an elementary classroom. In the comprehension phase they often focussed specifically on science knowledge with the purpose of designing a lesson centered on maintaining student interest and motivation. In transforming information or ideas for an elementary science lesson in the school setting, they had either relied on an activity trialled or discussed in the first-year science education subject, or from a lesson or worksheet supplied by their co-operating teacher during the school-experience part of their program.

The PBL program provided opportunities for participants to develop their pedagogical reasoning, and this was achieved through the questions asked in the problem and the journal, discussion in the group, and independent study. This chapter provides an overview of the pedagogical reasoning developed by members of the class group. A more detailed review of the pedagogical reasoning for two preservice teachers has been reported in Peterson and Treagust (1998).

Comprehension: From the discussion in the previous section, it is evident that preservice teachers were able to develop an understanding of the science content knowledge, and the curriculum knowledge associated with their individual topics.

Transformation: During the planning stage, members of this group considered the prior knowledge of elementary students, the importance of developing activities which were relevant to learning, considered the questions to be used in the classroom, and began to consider evaluation options as part of the planning process. When selecting activities for teaching, all preservice teachers were more focussed on the value of the activity to develop the science concepts relevant to the topic. ("I think, with regards to planning, I need to come up with some activities that are meaningful and not just included because they look good. This is very important in any lesson. The child may have a great time, but what did they learn?" [Tony/Journal-2/11])

Most preservice teachers discussed the questions that they would use to focus their learning on the relevant concepts. Students' prior knowledge of the topic based on questions asked in the problem scenario was used in this planning process. ("From the questions asked, it seems the children do not know much about their own eyes and how they work. Therefore, we've decided that first they need to look at the structure of the eye to be able to identify the parts". [Michelle/Journal-2/8].) Establishing the knowledge that was appropriate for elementary students to learn compared to their own understanding of the topic was an important aspect of the transformation stage.

The type of questions that preservice teachers planned to ask during the teaching phase was a focus of this transformation stage, with most participants discussing or listing the questions they would ask in their journal. For example, Michelle decided to use question such as "How are our eyes protected? Why are two eyes better than one? Why is it difficult to catch a ball with one eye closed?" [Michelle/Journal-2/8] for discussion and probing student understanding of the ideas, whereas Sally considered questions that could be asked more generally:

> I have mentally divided questions into 3 groups, and must make sure to spread these evenly throughout the lesson. Type 1: What do you know about ...? (Assessing children's prior knowledge and pre-conceptions) Type 2: What do you think is happening? Why is it occurring? (These will help the children to focus on the concept and will help me assess their developing understandings). Type 3: What if ...? (Applying knowledge to other situations) Fourthly, it is important to give the children the opportunity to ask their own questions. (Sally/Journal-2/10)

Recognizing appropriate learning experiences and prior learning of the students, and considering avenues for learning through the appropriate use of questions are all important aspects of the pedagogical reasoning process.

During this transformation stage, 15 students reported in their journals that they spent some time considering an appropriate explanation of the science ideas for elementary students, which initially helped them develop an understanding of the concepts. ("I have listened to myself explaining some elements of this unit and I must remember to keep the language simple, for example, light is bent rather than refracted. This has been a valuable unit for me as far as addressing this problem". [Sally/Journal-2/10].) For other preservice students, less confident in their ability to teach science, this process of developing explanations appeared to help them decide ways to discuss ideas with students.

The questions provided in the scenario encouraged preservice teachers to consider the evaluation strategies appropriate for their topic. Consistent with their experiences in secondary science education, the preservice teachers' initial views of an evaluation strategy was an end of unit test to establish what the students had learned. As a result of this PBL unit, ten preservice teachers provided journal descriptions of evaluation approaches that focussed on student learning through observing students when doing activities in class, listening to discussion between students, participating in discussions with students, and encouraging students to self-evaluate their progress on activities. Three preservice teachers were teacher-centered in their discussions and focussed on evaluating knowledge through paper-based tests. The remaining students combined a mix of these two approaches.

The transformation phase enabled preservice teachers both individually and in a group to begin considering issues in planning to teach a science topic. Even when the ideas varied between group members, the opportunity to explore, examine and defend various ideas in a group situation was important to developing the thinking associated with planning to teach science.

Instruction (peer teaching): Preservice teachers taught part of their topic to a small group of their peers ($n = 3$) and then spent some time discussing the teaching and the possible approaches for teaching the topic with elementary students. One of the main purposes of this component of the PBL program was to allow the preservice teachers to proceed to the next three stages of the reasoning process. When teaching their peers, instruction varied from an exploratory approach identifying the participants' understanding, through to an approach discussing how to implement this topic in the classroom. Often the less confident students used this latter approach.

Evaluation, reflection and new comprehension: These three pedagogical stages are discussed together because of the inter-relationship between them, and in discussions with preservice teachers it was evident that their views on evaluation influenced their reflections and consequently their new comprehension of a topic or teaching.

As part of their journals, preservice teachers reviewed the peer-teaching activity and the overall PBL unit. In particular they were asked to consider their understanding of the science in their topic, and their ability to plan, teach, and evaluate science activities. All preservice teachers provided commentaries of which two have been reported in detail in Peterson and Treagust (1998). For example, Tracey noted that "The activities worked well but if I had to do it again I would not explain how to make an electric circuit step by step but give them the material and let the students find out for themselves... because children would understand it better if they discover it" (Tracey, Journal, p. 20). This peer teaching experience had enabled Tracey to reconsider her teaching role and whether or not she should explain ideas or allow students to explore ideas with a group. Mandy wrote that she realized that careful selection of resources was important. She suggested that it "allows children to explore concepts" and that it was important to "listen to the children's responses in group discussion closely because this would provide [her] with greater insight into where the lesson was heading" (Mandy, Journal, p. 21).

Preservice teachers were often surprised at the low level of understanding of their peers as they taught a given topic. The peer teaching was useful in that from this context it highlighted the importance of finding out what the learner already knows as part of the teaching process. As an introduction to a PBL approach, the use of a semi-structured journal focussing on the pedagogical reasoning process enabled students to work in a self-directed and independent manner without needing a tutor to guide the process.

CONCLUSIONS AND IMPLICATIONS FOR TEACHER EDUCATION

In this chapter, we describe a problem-based approach for the teaching of science education. Working in small groups and without the assistance of a tutor, preservice teachers are required to develop their knowledge of science content, curriculum and learners. To assist in this development, a pedagogical reasoning framework was used to provide the preservice teachers with guiding questions and a structure when addressing their problem.

In this PBL program, preservice teachers believed the approach had enabled them to develop their knowledge of science content, curriculum, and learners, and to use of a reasoning process to make decisions about teaching and learning. The problem-based case scenarios enabled preservice teachers to develop their science content knowledge and to explore possible curriculum activities associated with their topic. Fifteen of the sample of 21 preservice teachers were able to consider the learners' prior knowledge in their planning of science activities. The remaining six members still were teacher-directed in their approach. All 21 participants were able to focus on the development of pedagogical reasoning by completing a semi-structured journal, which contained guiding questions to assist their thinking and reflective process. One of the important aspects of this PBL approach was that it allowed preservice teachers to be self-directed and have greater ownership of their learning and to explore areas where they perceived further understanding was needed. For this reason, the level of teacher control of learning shifted significantly to the students.

The scope of the PBL approach is unlimited for preservice and inservice science teacher education programs. It is only dependent on the development of relevant problem-base scenarios to meet the needs of the curriculum, and the participants' prior knowledge. For example, with the careful selection of scenarios, it is possible to consider a range of issues such as gender, assessment, and science and technology within a program. Problem-based scenarios can easily be developed as classroom situations provide the contexts for the development of scenarios. As students become more proficient in the approach the need to provide guiding questions in the problem scenario, and in the use of a semi-structured journal may not be as important. In addition, the complexity of the cases can also increase as students become more confident in using the approach.

One of the criticisms of a PBL approach is that the students take longer to complete the intended curriculum and do less when compared to a more traditional teaching approach because of the time spent in small groups. What must be considered in responding to this criticism is that PBL is not just about acquiring a body of knowledge but also developing the student's ability to reason, make logical connections between ideas, and acquire the skills to be more self-directed and lifelong learners.

REFERENCES

Barrows, H. S., & Tamblyn, R. M. (1980). *Problem-based learning—An approach to medical education.* New York: Springer.

Boud, D. (1985). Problem-based learning in perspective. In D. Boud (Ed.), *Problem-based learning in the education for the professions* (pp. 13-18). Sydney: HERDSA.

Boud, D., & Feletti, G. (Eds.). (1991). *The challenge of problem-based learning.* London: Kogan Page.

Chappell, C. S., & Hager, P. (1995) Problem-based learning and competency development. *Australian Journal of Teacher Education, 20,* 1–7.

Cleary, E. G. (1996). *Problem based learning in a large teaching format.* LEAP [On-line]. Available: http://web.acue.adelaide.edu.au/leap/focus/pbl/

Engel, C. E. (1991). Not just a method but a way learning. In D. Boud, & G. Feletti (Eds.), *The challenge of problem-based learning* (pp. 23-33). London: Kogan Page.

Grossman, P. L., Wilson, S. M., & Shulman, L. S. (1989). Teachers of substance: Subject matter knowledge for teaching. In M. Reynolds (Ed.), *The knowledge base for beginning teachers* (pp. 23-36). New York: Pergamon.

Heycox, K., & Bolzan, N. (1991). Applying problem-based learning in first-year social work. In D. Boud, & G. Feletti (Eds.), *The challenge of problem-based learning* (pp. 186-193). London: Kogan Page.

Kennedy, M. M. (1990). *A survey of recent literature on teachers' subject matter knowledge.* National Center for Research on Teacher Education, Michigan State University, East Lansing.

Kruger, C., & Summers, M. (1988). Primary school teachers' understanding of science concepts. *Journal of Education for Teaching, 14,* 83–95.

Lovie-Kitchin, J. (1991). Problem-based learning in optometry. In D. Boud, & G. Feletti (Eds.), *The challenge of problem-based learning* (pp. 194-202). London: Kogan Page.

Margetson, D. (1993). Understanding problem-based learning. *Educational Philosophy and Theory, 25,* 40-57.

Marsh, C. J. (1992). *Key concepts for understanding curriculum.* London: Falmer Press.

McDiarmid, G. W., Ball, D. L., & Anderson, C. W. (1989). Why staying one chapter ahead doesn't really work: Subject-specific pedagogy. In M. Reynolds (Ed.), *The knowledge base for beginning teachers* (pp. 193-206). New York: Pergamon.

Merriam, S. B. (1988). *Case study research in education—A qualitative approach.* California: Jossey Bass Inc.

National Academy of Sciences. (1995). *National science education standards* [On-line]. Available: http://www.nap.edu/nap/online/nses/.

Peterson, R., & Treagust, D. F. (1992). Primary pre-service teachers' pedagogical reasoning skills. *Research in Science Education, 22,* 323–330.

Peterson, R., & Treagust, D. F. (1995). Developing preservice teachers' pedagogical reasoning ability. *Research in Science Education, 25,* 291-305.

Peterson, R., & Treagust, D. F. (1998). Learning to teach primary science through problem-based learning. *Science Education, 82,* 215-237.

Reynolds, A. (1992a). Getting to the core of the apple: A theoretical view of the knowledge base of teaching. *Journal of Personnel Evaluation in Education, 6,* 41-55.

Reynolds, A. (1992b). What is competent beginning teaching? A review of the literature. *Review of Educational Research, 62,* 1-35.

Ryan, G., & Little, P. (1996). Bibliography. *Probe,* Newsletter of the Australian Problem-Based Learning Network. PROBLARC, University of Newcastle, NSW.

Shulman, L. S. (1986). Those who understand: Knowledge growth in teaching. *Educational Researcher, 15*(2), 4-14.

Shulman, L. S. (1987). Knowledge and teaching: Foundation of the new reform. *Harvard Educational Review, 57,* 1-22.

Shulman, L. S., & Sykes, G. (1986). *A national board for teaching? In search of a bold standard.* A paper prepared for the Task Force on Teaching as a Profession. Carnegie Forum on Education and the Economy, Stanford, California.

Speedy, G. (Ed.). (1989). *Discipline review of teacher education in mathematics and science.* Canberra: Australian Government Printing Service.

Stofflett, R. T., & Stoddart, T. (1993). The ability to understand and use conceptual change pedagogy as a function of prior content learning experience. *Journal of Research in Science Teaching, 31,* 31-51.

Treagust, D. F., Duit, R., & Fraser, B. J. (Eds). (1996). *Improving teaching and learning in science and mathematics.* New York: Teachers College Press.

Wilson, S. M., Shulman, L. S., & Richert, A. (1987). "150 different ways of knowing": Representations of knowledge in teaching. In J. Calderhead (Ed.), *Exploring teacher thinking* (pp. 104-124). Sussex, England: Holt, Rinehart & Winston.

Woods, D. R. (1994). *Problem-based learning: How to gain the most from PBL.* Waterdown, Ontario: Donald R. Woods.

LINKING SCHOOLS AND UNIVERSITIES IN PARTNERSHIP FOR SCIENCE TEACHER PREPARATION

Marcia K. Fetters[1] & Paul Vellom[2]
[1]*Western Michigan University*, [2]*Ohio State University*

INTRODUCTION

Historically, schools and universities have worked together in relationships characterized by a set of more or less informal agreements that provided contexts for the preparation of new science teachers and the growth of knowledge about science teaching and learning. Negotiations for this type of relationship between schools and universities were usually undertaken on an as-needed basis, and determined the parameters of the relationship for a specific duration of time. A key aspect of many, if not most, of these relationships was the university preparation of pre-service teachers who needed some practical experience in teaching before being given full responsibility for a classroom of students. Within these relationships, experienced teachers participated as 'cooperating teachers' for two main reasons: they felt a debt to their own original cooperating teacher or preparation program and wanted to provide opportunities for others to join the profession; and they saw themselves as having something to contribute to teacher preparation. Another historically common type of school-university relationship typically emerged when schools asked for advice and support from a nearby university, and the university set up one or more research projects in these schools. In these cases, participating schools often have had specific needs in teacher development or for expert consulting related to some kind of improvement project. Meanwhile, universities have seen the schools as likely sites for investigating problems and successes in teaching and learning. More recently, some school-university relationships have shifted from the earlier "individual needs" based model to more robust forms that engender mutual benefits. While recognizing a wide range of available models, this chapter focuses on the work of the Holmes Group, and examines this model in respect to the preparation of science teachers. In particular, we address preparation of teachers in concert with visions of teaching captured in current science education reform documents and in concert with a constructivist view of classroom learning.

THE HOLMES GROUP AND FOUR GUIDING PRINCIPLES

In the 1980's the Holmes Group, a consortium of over 100 education research institutions, was incorporated and began working toward making previously short-term

D.R. Lavoie and W.-M. Roth (eds.), Models of Science Teacher Preparation, 67–88.

or informal relationships between schools and universities formal. Beyond just a formalization of previous relationships, the consortium strove to describe how partnerships between schools and universities required rethinking the structure and work of both institutions. The formalization of relationships brought with it some real strengths and benefits for all parties involved. Over time, the Holmes Group developed a set of three documents to help guide its groundbreaking work: *Tomorrow's Teachers*, *Tomorrow's Schools* and *Tomorrow's Schools of Education* (The Holmes Group, 1986, 1990, 1995). 'Professional Development School' is the term coined by the Holmes Group to describe the organization, structure and activities of a school in partnership with a university.

In developing these guiding documents the Holmes Group (1990) identified four principles to guide the work of the members of a partnership and lead to the institutionalization of the partnership. These include reciprocity (mutual exchange and benefit between research and practice), experimentation (willingness to try new forms of practice and structure), systematic inquiry (new ideas are subject to careful study and validation), and student diversity (commitment to the development of teaching strategies for a broad range of children with different backgrounds, abilities, and learning styles). On the basis of these four principles, traditional school-university relationships (previously based on individual needs) were transformed into the Professional Development School model, in which a partnership is based on shared goals between schools and universities.

ESTABLISHING A PARTNERSHIP—A MODEL

The Professional Development School (PDS) model assumes three main foci: preservice teacher education, in-service professional development, and inquiry into teaching and learning. This partnership model differs from more traditional relationships between schools and teacher preparation institutions in important ways (see Figure 1).

While literally dozens of examples of school-university partnerships can be found in popular and research literature today, only a minor fraction of these partnerships can be traced to an initial set of shared goals between partner schools and universities. (For another model of this type see Barufaldi and Reinhartz, this volume.) Notable among these are partnerships found within the National Network for Educational Renewal (1999) and the Coalition of Essential Schools (1998). While each of these long-term projects has its own vision for school reform, each also reflects the three facets of the PDS model: preservice teacher education, inservice professional development, and sustained inquiry (Cushman, 1993; NNER, 1999).

In these and other examples that focus on urban schools in the United States (e.g., Sobel, French, & Filbin, 1998) and on partnerships in European contexts (e.g., Day, 1998), the four Holmes Group principles of reciprocity, experimentation, systematic inquiry, and student diversity are defined and elaborated in various ways. Research studies portray a robust range of partnerships, each designed to meet contextual demands while reflecting these principles. We see them as an expected and positive outcome of the work of the Holmes Group and many other efforts to revitalize schools and rethink teaching practices. In particular, the wide range of partnerships has positive implications for teacher preparation, which has traditionally fallen prey to 'turf wars' between school-based and university-based practitioners.

Traditional School-University Relationships

School Needs	*University Needs*
1. Professional development: courses and workshops for teachers. 2. Staffing needs: Identification of promising new teachers to fill vacancies and develop new programs.	1. Sites for student teaching and methods students, to work with experienced teachers to serve as cooperating teachers. 2. Sites for educational research

Professional Development Schools

Shared Goals and Activities		
Pre-service Teacher Education	In-service Professional Development	Inquiry into Teaching and Learning
Examples: 1. Prepare effective new teachers for today's diverse student populations 2. Establish and support mentoring relationships as a centerpiece of pre-service teacher education Develop examples of teaching practice and productive methods that can catalyze learning in pre-service teacher education	Examples: 1. Link theory, current research, and best practices to better serve all students 2. Collaboratively set and work towards site goals that will enable better teaching in the short run, and longer-term development and retention of career teachers who are committed to continuous renewal of their own teaching practices and those of colleagues	Examples: 1. Systematically study examples of teaching and learning in order to meet all students' needs 2. Together, develop and test new models for curriculum, instruction, and assessment 3. Study the structures of schools that impact teaching and learning, and modify and test new structures

Figure 1: Two kinds of school-university partnerships

The Holmes Group model for Professional Development Schools was written as a broad model encompassing all grade levels and all disciplines. Figure 2 shows that the organizing principles of the Holmes Group and science education reform efforts *National Science Education Standards* (NRC, 1996) and *Blueprints for Reform* (AAAS, 1998) are both compatible and mutually transforming. In this table, we illustrate in more concrete terms how deeply resonant these models and principles are with other significant efforts to reform teaching and learning in science. Essentially, with such strong resonance, we bring into focus significant writings on the wider range of efforts in Professional Development Schools (e.g., Levine & Trachtman, 1997). This still recognizes the value and importance of many other models which impact teacher preparation (e.g., the 'teacher-

Holmes Partnership	*National Science Education Standards*	*Blueprints for Teacher Education*
Teaching and learning for understanding. All the school's students participate seriously in the kind of learning that allows you to go on learning for a lifetime. This may well require a radical revision of the school's curriculum and instruction.	*Images of science content.* The actions of teachers are deeply influenced by their perceptions of science as an enterprise and as a subject to be taught and learned.	*Convey a broad vision of scientific literacy.* Teacher education should produce science teachers who are committed to increasing understanding of connections between science, mathematics, and technology as well as understanding of the social, historical, and philosophical contexts of scientific knowledge.
Creating a learning community. The ambitious kind of teaching and learning we hope for will take place in a sustained way for large numbers of children only if classrooms and schools are thoughtfully organized as communities of learning.	*Learning is social in nature.* Student understanding is actively constructed through social processes.	*Teaching as a social enterprise.* Schools should promote and encourage team teaching. By promoting team teaching and other networking opportunities teachers develop reflective practices that leads to increased science understanding.
Teaching and learning for understanding for everybody's children. A major commitment of the Professional Development School will be overcoming the educational and social barriers raised by an unequal society.	*Recognize and respond to student diversity and encourage all students to participate fully in science learning.* In determining the specific science content and activities that make up a curriculum teachers consider the students who will be learning the science.	*Enable teachers to teach science to all Americans.* New teachers must be sensitive to student differences and understand how individual characteristics and special needs can affect engagement and achievement in science. New teachers need experiences in schools that model how the diverse needs of students can be addressed.

Continuing learning by teachers, teacher educators, and administrators. In the Professional Development School, adults are expected to go on learning too.	*Professional development is continuous, lifelong process.* Professional development for a teacher of science is a continuous, lifelong process.	*Emphasize the profession of teaching.* The primary purpose of teacher education should be to develop in teachers the attitudes, knowledge, and understanding that enable them to apply theories; and principles in devising strategies and classroom activities that are responsive to the needs and backgrounds of students. Teacher education must be redefined as a career-long endeavor.
Thoughtful long-term inquiry into teaching and learning. This is essential to the professional lives of teachers, administrators, and teacher educators. The Professional Development School faculty working as partners will promote reflection and research on practice as a central aspect of the school.	*Ongoing professional development.* The conventional view of professional development for teachers must shift from technical training for specific skills to opportunities for intellectual & professional growth. Professional development occurs in many more ways than delivery of information in the typical university course, institute, or teacher workshop. Teachers could conduct classroom-based research, or participate in research at a scientific laboratory.	*Enhance teacher learning.* Teacher education programs must offer prospective teacher opportunities to observe, experience, and participate in activities that emphasize student-centered and hands-on learning. Prospective teachers should use the knowledge from educational research to examine the practice of teaching in new ways, and need opportunities to discuss classroom experiences in light of formal knowledge about teaching and learning.
Inventing a new institution. The Professional Development School will need to devise for itself a different kind of organizational structure, supported over time by enduring alliances for all the institutions with a stake in better professional preparation for school faculty.	*Professional development opportunities need to be clearly and appropriately connected to teachers' work* . Whenever possible, the professional development of teachers should occur in the contexts where the teachers' understanding and abilities will be used.	*Implement reform.* Professional organizations, administrators in higher education, and administrators in K-12 education are the keys to the success of any reform, and should be included in professional development.

Figure 2. Illustration of linkages between Holmes principles and recent Science Education reform efforts

leader' model as found in Wasley, 1991) particularly in science. Below, we choose to focus on the Holmes-PDS model as an indicator example. We thereby provide a discussion of partnerships for teacher preparation, including partnership parameters that must be addressed in every partnership. We believe that such a discussion would be much less meaningful if expressed in the general terms required to include many partnership models. We leave it to the reader to make appropriate adjustments and applications, as they are warranted.

Developing a school-university partnership requires attention to three main areas: defining the partnership, people resources, and maintaining and advancing the work over time. In the next section, these fundamental areas of endeavor are described in greater detail, and example vignettes are provided.

DEVELOPING A PARTNERSHIP

Defining a Partnership

There is no single model that will work for all schools and universities in partnership. Each partnership will evolve in its own ways over time, shaped by available resources and identified needs. Within partnerships, there are four areas of concern that that institutions and individuals should consider in structuring school-university collaborations. These include identity, roles in the partnership, goals of the partnership, and necessary resources.

Developing an identity involves clearly defining the nature of the partnership. This includes identifying long-term and short-term goals and articulating the purposes of the partnership. If the partnership is to be ongoing and sustainable over time, thought and consideration must be given to plans for initial work in the partnership and strategies for addressing changing needs and goals of the partnership over time. Roles in the partnership become important once preliminary goals and purposes of the partnership are established. Using the strengths and skills of each partner as a basis for determining roles and responsibilities, the next step is to identify individuals who will be involved and their responsibilities. Over time, all partnerships must address how to bring new members into the partnership. When using school-university partnerships as an important part of teacher education program all members of the partnership must identify significant ways to engage pre-service teachers in the partnership.

Goals of the partnership should recognize that members bring their agendas and priorities to the work; for this reason, developing a new, shared set of goals is essential. Even so, individuals will experience different levels of ownership in various partnership activities. As the partnership grows and evolves over time, the work, goals, and priorities of each of the members of the partnership must be actively balanced across three activity areas: in-service professional development, pre-service teacher education, and inquiry. Decisions must be made to determine who will take leadership roles, to balance priorities, and to select pathways to move forward on the partnership agenda. Goals should include plans for evaluating the progress of the partnership. Evaluation should viewed as a continual process that examines short- and long-term goals, and provides information feedback loops for setting or modifying partnership goals.

Resources must be an early consideration whenever a new partnership is formed. Members of the partnership together identify the resources necessary to sustain the partnership, which may include individuals with skills or expertise, funding to pay for materials, funding for support personnel, dedicated work space, and time within the job descriptions of each of the participants.

School-university partnerships are often inaugurated with one or two main purposes or common goals. Over time, goals requiring deeper commitment emerge. The following two vignettes illustrate a partnership in it's earliest stages. The first describes a school district in need of raising student test scores on state-administered achievement tests, a partnership in which the initial needs come from the school and district. The second vignette shows how, over time, the needs of the university are brought to the partnership. These two vignettes paint a picture of some of the challenges and issues that beginning partnerships commonly face.

Vignette 1: Curriculum and School Reform in Light of State Science Tests

The Superintendent of Springvale, Sarah Steward, has called an emergency meeting. The results of the state wide assessment tests are in. Only 35% of the students in the Springvale district have met the minimum requirements for their grade level. Sarah knows that these low scores are not acceptable in her district: they are lower than last year's scores. The problem appears to be systemic, especially at the middle school and high school levels. Approximately 50% of the students at the elementary level are meeting their goals, but the percentage of students passing at the middle school level drops significantly, with even fewer passing at the high school level. Sarah also knows that, in the near future, the high school test will probably be tied to a state-endorsed diploma.

Superintendent Steward formed a committee for each subject area tested, comprised of ten teachers across all grade levels. The science committee was made up of one teacher from each of the of the four elementary schools in the district, a teacher from each of the three grade levels at the middle school, and three teachers from the high school representing biology, chemistry and physics. During the first couple of committee meetings, the teachers decided that they would focus on four areas that they could explore across grade areas: Interdisciplinary science, thematic cross-subject teaching, inclusion settings, and portfolio assessment. (On these special issues see the chapters by Yager *et al.* [interdisciplinary science, cross-subject teaching], Rennie [gender], Thomson *et al.* [multiculturalism], and Stein [portfolios].) Discussion led to acknowledging that they both needed and wanted some outside help as they worked on revising and further implementing the existing curriculum reform efforts, and documenting the successes they were already seeing.

A few teachers on the committee had previously worked with faculty from Green Valley University when they had agreed to have student teachers in their classrooms. Yet, some committee members were leery of working with the university.

They did not want university personnel to come in and take over. They wanted individuals from the university to work with them as equals who had similar interests, but also new ideas and different skills. To promote such a relationship they decided to set goals and then try to identify resources (including individuals) for each. This then enabled them to develop an action plan, a staffing plan, and a timeline for working on each goal.

Lessons from vignette 1

In this early example, school-university collaboration occurred first in committee, as teachers asked for university support and the planning team did its work. Partnerships of this type should have open conversations about what is needed from each participant, and how the work will evolve over time. This vignette also shows the importance of developing specific plans for how partnership activities will be accomplished, and by whom. Partnerships often stumble (and many fail) when expectations are not made clear and members do not deliver on promises, real or perceived. This vignette is idealized. No mention is made of school partners' compensation for additional work, or university partners' consulting fees. Instead, assumptions about the importance of these kinds of conversations, and the level of collegiality were required allowing individuals to negotiate common goals and work plans.

Barriers and Recommendations

Time can be one of the biggest stumbling blocks to a partnership. Individuals who are usually drawn to this kind of work often have full and busy lives. Scheduling "just one more" activity can become a real burden. Even adding in the planning and consulting time to work with pre-service science teachers can seem like a burden. Both schools and universities are driven by set schedules and these schedules are not always compatible. As time cannot usually be found, the formation of a partnership means rearranging priorities so that the partnership does not become one more burden to already busy lives. Some ideas include (a) arranging for common planning or activity times, which allow science teachers to coordinate curricula and look for ways of integrating the sciences and bringing in community-based resources; (b) scheduling working breakfasts or lunches, providing opportunities to brainstorm options, and to evaluate the feasibility of more innovative ways of teaching science; (c) using in-service time for partnership activities; (d) providing weekly after-school seminars to expand teachers' knowledge of science and science pedagogy; and (e) considering monthly Saturday meetings to provide an expanded time frame for complex laboratory or learning activities. The energy to engage in conversation and debate is often greater than in after-school meetings. Similar to in-service days, more teachers can be involved.

Workspace and related resources cannot be overlooked. Partnership activities most often require some dedicated space, with additional space that can accommodate large group meetings and breakout groups. Basic resources such as an easel with newsprint, a chalkboard or white board and adequate electrical outlets for laptop computers and other media devices are necessary. Partnerships also need dedicated space to store materials so all that partners have easy access to them. Places where partners can store personal belongings and where they can work together also help partners feel like partners, not just visitors. Space at both the school and university is needed. The necessity of these resources to the success of the partnership make the initial time and energy investments they require worthwhile. Our recommendations included (a) rearranging class schedules so that some rooms are used every period each day, freeing up one or more classrooms for partnership meetings and work; (b) seeking easy access to the internet and world wide web, which can facilitate partnership work including arranging e-mail and university library privileges for school partners; (c) creating access to multimedia equipment for all partners; assembling a storage area with file cabinets or bookshelves, including secure storage for technology; (d) providing all partners with keys to access work and storage spaces at the school and university; and (e) providing mailboxes and a mechanism for delivering memos and materials between sites.

Physical distance can be a barrier, and partnerships cannot be based merely on proximity. Individuals must share some common goals or beliefs for a partnership to be successful. Even so, when a school and a university are located a great distance apart, major challenges can be expected. Drive time, parking, and scheduling are all issues that will significantly impact the work of the partnership. Recommendations include (a) moving some university courses to the school site if space permits, travel, and providing transportation for pre-service teachers; (b) arranging partnership activities in the early morning or late afternoon so that partners do not have to run back and forth between sites; (c) holding some partnership activities at a meeting place halfway between sites; (d) providing parking passes or dedicated parking spaces for partners at both sites to enable commuting between sites; (e) blocking out larger time periods for partnership activities to reduce the number of meetings; and (f) establishing and providing access to electronic communication systems.

Vignette 2: Mentoring Novice Teachers

As the team at Springvale worked together, other topics entered their conversations. University partners began talking about some of the things they wanted pre-service teachers to experience and learn about when they came into Springvale schools—real examples of teachers using some of the strategies and theories addressed in university classes. They also needed help in identifying safe places for pre-service teachers to try these strategies in a supportive environment. In particular, they wanted pre-service teachers to see inexperienced teachers using strategies such as teaching as inquiry, problem-based science inquiry; cooperative learning and group work, writing across the curriculum, and authentic science assessment.

University partners also wanted pre-service teachers to have direct experience in planning, setting up, and tearing down laboratory activities. Laboratory safety and equipment and material management were topics already included in methods courses, but required attention in a real classroom setting. The question, "How do I set up and organize laboratory work?" seemed central to their goals and required consideration in authentic school settings working with real students.

The school-based partners also had history-based concerns about field experiences and student teaching, mainly focusing on the lack of communication between teachers in the field and the university programs and personnel. In the past, pre-service teachers would just show up in classrooms, without clear information about what they should be doing, and not being able to articulate their tasks or assignments to the cooperating teacher. School-based partners were often unsure of their roles when working with pre-service teachers. Pre-service teachers appeared to lack skills in student management and in pacing a lesson. When teaching lessons, pre-service teachers often finished before class ended or were caught by surprise when the bell rang.

School-based partners also had concerns about pre-service teachers' science content knowledge, particularly about the kinds of knowledge and understanding that the students could and could not demonstrate. Across grade levels in today's schools there is an increased focus on the integration of science disciplines and the role of science in everyday life. However, science as it is taught at most universities is neither integrated nor are daily applications of science emphasized.

Another issue was raised by university partners: pre-service teachers often complained about having to write out formal lesson plans, especially after their experiences in schools. In the field they saw that teachers usually used a plan book that was divided into boxes and a whole day's activity was captured in one box, with only 15–20 words describing the whole day. Pre-service teachers judged the situation at face value, and believed that the university's emphasis on planning did not match the true demands of the profession.

As these issues emerged and became a significant part of the conversation, school- and university-based partners decided to form a committee to examine field experiences and student teaching. Figure 3 shows how they decided to work on the goals and address the concerns of each set of partners as they began conversations about building a better and more systemic relationship around pre-service education.

Lessons from Vignette 2

Preservice teacher education is an area in which university partners have greater ownership, but school partner support is essential. As the partnership matured, the initial responsibilities exhibited in Figure 3 evolved and became more specific for science instruction. Among the issues raised by university partners was lesson planning. Specifically, the university partners wanted the mentors teachers to understand the importance of lesson planning in the preservice program, and to enlist their support in requiring detailed lesson plans of pre-service teachers. The university partners saw the brief written plans of these experienced teachers as the product of past experience and well-developed repertoires of teaching strategies and approaches, things that the pre-service teachers lacked. Hence, detailed plans were seen as a path to good teaching which would eventually enable the pre-service teachers to understand that the mentors did a lot of the detail work in their heads. While this issue was of critical importance to the university partners, it did not enter into their negotiation of the initial work plan for the partnership. Instead, it was seen as an issue that would best be negotiated later, during the time that placements and expectations for pre-service teachers were being worked out.

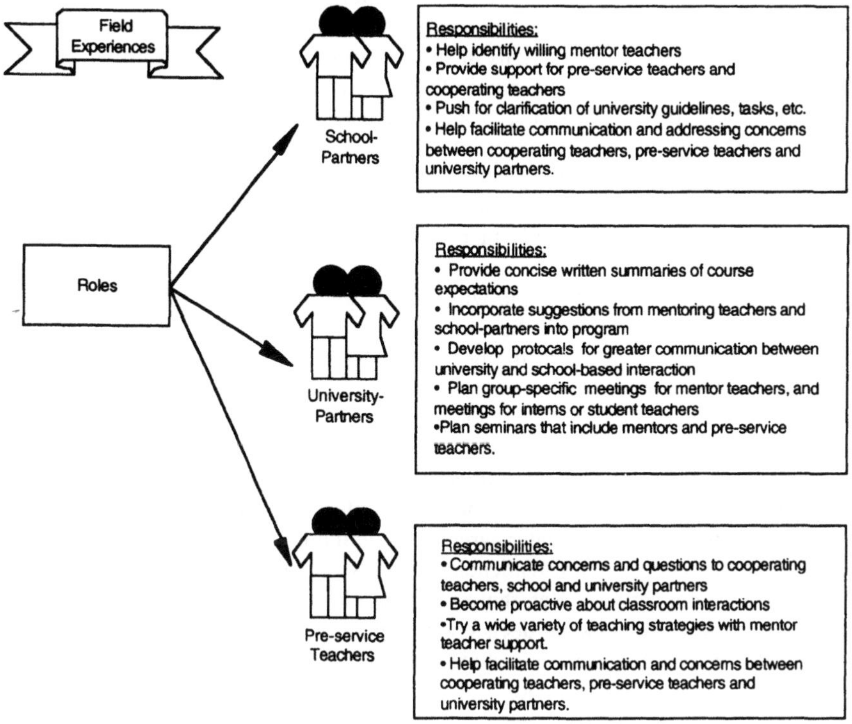

Figure 3: One example of a work plan for field experience in a school-university partnership

Barriers and Recommendations

Since field experiences in classrooms are an important part of school-university partnerships, getting buy-in from a majority of the members of a school science department is desirable. This allows a substantial number of students to be placed in a school, and school partners are enabled to play significant roles in the teacher preparation program. From the university perspective, this means finding out whether science teachers at a school really want to join the partnership, and identifying and securing commitments from individuals at the university who are interested in partnership activities. Sometimes the biggest barrier to getting people involved in a project is a lack of personal communication and a personal invitation. Ownership of the process and of partnership activities must be shared among partners if the partnership is to succeed and last. Some ideas include the following: (a) distribute a flyer or memo introducing the project and it's benefits be followed by personal communication and encouragement and for science and other departments and provide all members with an updated contact list of people involved in the partnership; (b) schedule informal conversations at lunchtime or in the early morning so that people can gather and talk about possibilities; (c) plan in-service sessions in which members of other partnerships visit to share the types of work they have been doing; (d) visit other science departments who are part of school-university partnerships; (e) provide as much salient information as possible so that partners can consider options, envision changes in their work settings, and start to take ownership in partnership activities; and (f) honestly talk about the nature of the work and the commitment requires.

In any partnership, some people will initially get involved and will later decide that the partnership does not fit their life goals or priorities. Other individuals who were hesitant to get involved at the beginning will see good things happening and want to join. Over time, some members of the partnership will move on to other positions in the school or university and new people will be hired to take over their duties. Every partnership will face the challenges of bringing new people into the partnership, which include catching them up on past work and empowering them to influence future work. Partnerships change, people come and go. This is a natural evolution and partners should enter agreements with this in mind. When a key player leaves the partnership, natural tendencies are to view this event as a failure on the part of one or more partners. Yet, those same people may opt out for a few years and then rejoin. These are personal decisions and may or may not reflect their view of the partnership. Enabling actions include keeping a written history of the partnership that documents the decisions made and logs previous work. Set aside time in regularly schedule meetings for newer members to suggest projects or goals. Actively seek their input by being open to new ideas and directions for future work and by being willing to renegotiate roles, goals and priorities. Do not make assumptions about new member roles. Provide new members with leadership opportunities as their talents and interests become clear.

SUSTAINING PARTNERSHIPS

Whenever groups of people from two or more institutions decide to form a partnership, established roles must be modified and new roles formed. Individuals who had a strong sense of purpose coupled to well-defined roles in previous arrangements may now find their roles shifting. Partners must select and focus on high-priority activities and responsibilities as previous roles evolve and change and as new skills and knowledge are developed.

A school-university partnership often calls for new roles beyond the traditional cooperating teacher, student teacher, and university supervisor (faculty member). These may include liaisons, mentor teachers, adjunct faculty, project directors, program coordinators, clinical educators, and documentors. Just as definitions for these roles will vary from partnership to partnership, so will the activities that each role encompasses. These are an important point of negotiation that must take place early, and that should be revisited regularly over the life of the partnership.

Partnerships almost always call for the reallocation of many kinds of resources. Negotiations between partners should address funding issues and other resources, and should be periodically reviewed so that when priorities change, the allocation of resources also changes. New resources may also be needed as new priorities and interests emerge. The definition of 'resource' must include the important resource of time. A school-university partnership calls for both schools and universities to examine their schedules so that rigid schedules get some flexibility built into them, and members become willing to be accountable for their time to their partners. A study of resources must include examination of reward structures in both schools and universities. If partnership activities are not recognized within the merit systems of partner institutions, sustaining the work of the partnership over extended periods of time will be difficult.

Documentation of partnership activities is of vital importance for the ongoing work of the partnership. Documentation provides a record of what work has been done, and by whom. It describes those things that have worked well and those things that were not as successful, providing a basis for future decisions and preventing a duplication of failed efforts. A detailed accounting of how decision-making and problem-solving processes have worked can be useful to new members as they join the partnership. Strong documentation accounts for the contributions of all partners and demonstrates equity, and often furthers inquiry processes as they develop within the partnership.

Behind the surface issues found in jointly held goals and related responsibilities, school-university partnerships always encompass common ground in both partners' interests in improving teaching and learning. Once a work plan and goals are established, individuals in each institution must commit to crossing traditional boundaries in order to make the partnership viable, and to realize its full potential. Vignettes 3 and 4 below illustrate how the Springvale-Green Valley partnership evolved to include crossing boundaries in both directions. In each case, players took significant risks to 'go against the grain' in order to effect change in their institutions.

Vignette 3: University Partners Working day-to-day and intensively in Schools

A major goal of a Professional Development School, as outlined by the Holmes Group (1986) is for university faculty and graduate student partners to take on active instructional roles in school settings. It is one thing for university partners to talk about different teaching styles and goals of science education reform, and it is quite another thing to model them for pre-service teachers and school partners! (On the topic of such collaboration, see Roth, this volume.)

As the work plan in Springvale District emerged, both school and university partners realized that the partnership was being constrained by their traditional ideas of school turf—that teaching and curriculum were the school's purview, and that the university partners had neither interest nor time to work in these areas. Yet, paradoxically, the university partners found many of the teachers hungry for collaboration in these same areas. Before long, three notable roles evolved for university faculty and graduate students working in Springvale schools: reassigned-time teacher, team teacher, and documentor/evaluator.

Reassigned-time teacher. Mr. Ewing and Mr. Lawrence taught Biology at Springvale High, where the science teachers had agreed to work together on redesigning curricula to better meet the requirements of the state tests. In order to provide time to support this curricular redevelopment, each was reassigned from teaching one class per day for a semester. During the summer prior to the year in which this time was available, Springvale's principal had begun negotiating directly with professors at Green Valley to line up an experienced Biology teacher for these two periods each day. Negotiations were made easier by adequate funding for the reassigned-time projects, supplied by the district. Still, issues of qualifications, expectations of each partner for the reassigned teacher's duties, topical expertise, and many issues related to roles confronted the partners as they made these arrangements. Finally, mere weeks before school was to begin, a graduate student with considerable teaching experience in biology was identified and contracted to fill the position. The strengths of this arrangement for both the university and the school lay in clear sets of roles and expectations, and the additional resources provided to impact teaching and curriculum.

In another model that provided reassigned time for a classroom teacher, Dr. Markum, the secondary science methods instructor, "borrowed" two of Ms. Charles' biology classes for two weeks. Dr. Markum, with the assistance of the students in his methods class, taught cellular division to two of Ms. Charles' classes. Ms. Charles was able to use this time to develop effective ways of using the new calculator-based probes that the district had just purchased. At the same time, she was able to monitor the class, enabling a smooth transition at the end of this time period.

Other collaborations were forged at several elementary schools in the district, where university faculty and graduate students took varied roles in instruction. Time commitments for these university partners spanned from a few days up to a semester. The enabling factor for these university partners was that some or all of their university responsibilities were designated as PDS work. Without this designation and recognition of time commitment, balancing their other responsibilities would have been difficult. Other faculty at Green Valley used sabbatical time or wrote grants to cover the cost of hiring part-time faculty or graduate students to take on some of their university duties, so that they would have time for intensive work in schools. Eventually, as the partnership between Springvale and Green Valley grew, university partners taught in many middle and high school classrooms for varying periods of time, exploring new curricular approaches.

Team teacher. At Springvale Middle School, teachers had been working in grade-level teams for several years. Each team of three teachers shared a set of about 80 students for mathematics, science, language arts, and social studies instruction. When the partnership emerged as a possibility, one middle school team became excited about the possibility of having a university person—they preferred a graduate student—to work with them on new teaching techniques, and to help them work towards a performance-based assessment system. This arrangement did not provide reassigned time for projects but an important resource person to bounce off ideas and to help facilitate and monitor student progress and activities. This kind of partnership served to reduce the isolation that these teachers felt while teaching, because the partners shared responsibility for the students. At one point during the year, the graduate student took a lead role in teaching, allowing one teacher to step into a supporting role. This allowed the teacher to examine some aspects of her students' work that she had previously missed. In teaming arrangements, then, partners can flexibly work out roles for teaching and support. Here again, the attitude of shared inquiry and a willingness to take risks are enablers.

Documentor/Evaluator. In every Springvale school in which university partners worked, roles related to documenting and evaluating the work of the partnership emerged. Some documentors took on instructional roles as well. In Springvale, each documentor worked with a teacher or team of teachers to help document teaching strategies and changes in the curriculum.

The day-to-day demands of teaching make it very difficult for most teachers to find time to reflect and write about their teaching. In Springvale, the university partners helped to capture—in writing or on videotape or audiotape—episodes of teaching and learning. This documentation was then used in a wide variety of ways. Teachers used it to share what they were doing in their classrooms with the community. It also demonstrated good practice as the district sought state and national recognition for effective schools. And, the information was analyzed by both school and university partners for various ends: to seek funding to implement a new curriculum and for research on teacher effectiveness and student learning.

Vignette 4: Schools Take Roles in the University

As the partnership with Springvale Schools developed and the university partners became more and more active in school buildings, the faculty at Green Valley University recognized opportunities to improve their teacher education programs. While they knew that some faculty colleagues would disapprove, their shared inquiry and openness to their school partners' ideas warranted working out new roles for these colleagues in preservice teacher education programs. Initially, they identified four different roles that school personnel could take on in interacting with a group of teachers: (1) team teaching in teacher education courses, (2) mentoring, (3) assisting in science-specific student teaching seminars, and (4) helping to establish and maintain student teaching support networks across disciplines.

Team-teaching in teacher-education courses. Dr. Norton was a faculty member involved in the Springvale partnership from the very beginning. His university responsibilities were mainly in teacher education, working with pre-service teachers, in-service teachers enrolled in the Masters' degree program, and doctoral students in science education. Initially he had a number of strong doubts about the partnership, but over time Dr. Norton came to see the teachers he met a Springvale High and Springvale Middle School as thoughtful practitioners who understood the realities of public school teaching better than he did. Over the course of the one-year preservice program, he invited several of these teachers to be guest lecturers in preservice courses. On one occasion, a teacher was well known for her use of writing to learn science. In the preservice program, this topic would usually be addressed in methods and issues courses and during student teaching. This teacher was asked to teach one to two sessions about the role of writing as a way of helping students learn in each of these courses. Another school partner became a team teacher for the university's science methods course, and later brought a panel of three other teachers to speak to the student teaching seminars on classroom management. Some of these courses met in school settings during the teacher's planning periods or after school. On occasion, the university partners swapped with a teacher, providing instruction for the school partner's classes when university courses were held during school hours.

Mentoring pre-service and first-year teachers. At Green Valley as nationwide, induction-year teaching and supporting beginning teachers had recently become a priority in science education, mainly to stem early-career attrition, but also to encourage growth as a regular part of the profession. The new view was that learning about the complexities of teaching in public schools, and in particular, strategies for surviving the many demands of teaching, could no longer be left for novice teachers to pick up on their own. At the University's urging, the Springvale District agreed to explore mentoring relationships to support new teachers, with a goal of implementing a formalized mentoring system in one or two years. The university and district were able to access several models for this type of work: the designs of Pathwise, PRAXIS and INTASC formalize this process. Together, they learned that Toledo Public Schools, for instance, had developed a nationally recognized program in which some master teachers' teaching duties were reassigned for a two-year period. These master teachers worked with beginning teachers in the school district to support and evaluate new teacher progress. They also learned that in the past, some of these master teachers worked closely with a university partner to provide support to these new teachers.

Assisting in science-specific student teaching seminars. Green Valley University coupled each field experience, including the student teaching experience, with a seminar in which students explored classroom issues in depth and built on each others' experiences for their own growth in the profession. Just as school partners had become active participants in other university courses, they became co-responsible for the student teaching seminars by suggesting important issues and providing resources for studying them, by meeting with small groups of students on occasion, and by appearing as guest experts or in role-playing situations. In this way, the seminars became more closely linked to the pre-service teachers' in-school experiences. As time passed, one of the instructors at Green Valley required that preservice teachers "have a conversation with your mentor teacher about..." an assigned topic as a regular, weekly assignment. Summaries of these conversations and reflections on their value became a mainstay of the seminar.

Establishing and maintaining a student-teaching support network. School-based partners in the Springvale-Green Valley partnership played an additional, important role in enculturating and supporting student teachers. This kind of support occurred both within science departments at their own grade levels, and working across disciplines and grade levels. When school partners had reassigned time, they were able to facilitate conversations between student teachers and university partners and help student teachers and cooperating teachers recognize some of the universal dilemmas that all student teachers face. The student teachers tended to hold these teachers' views and recommendations in higher regard than those of their university contacts.

Barriers and Recommendations

Conflicting agendas. Whenever two or more people work together, there will be times when individual agendas compete. Within partnerships, the question is not "Will this become an issue?" but rather "When will this become an issue?" How conflicting agendas are managed can lead to the partnership strengthening over time, or cause the partnership to dissolve. Our recommendations include the following four items. First, actively listen to each other during discussion and identify conflicts early. For example, quality science education for the K-12 students will always be a top priority for school partners. While this is also a major concern for university partners, preparing science teachers and science or science education research is very important. Teacher educators will often have driving questions about how children learn certain science concepts, and while the school partners are interested in this, they also face the time constraints of a district-mandated curriculum. Often, these agendas will come into conflict. Finding a way to value each agenda and establish a balance is essential. Second, seek out compromises. If partners wait thinking they will go away, they most certainly will grow into major issues that will require more time and energy. Often a decision about division of the work among partners will fail to take into the consideration how linked and connected various projects are, and how the work depends on a majority of the partners. Third, develop a consensus plan or protocol for voting on issues that maintains the integrity of the partnership by valuing its diversity. Once the group has made a decision about a direction, goal, or procedure, all members of the group must be encouraged to actively support the decision. While individuals may not agree with every decision, all partners must still be able support it rather than work against it or withdraw. Fourth, be flexible, and willing to take risks in considering new ideas about teaching science, and the new roles they suggest for all kinds of partners.

Science as elitist. No matter how strongly science and math teachers deny it, in most schools they are viewed as elitists by many of their colleagues in teaching. Science teachers in particular are often viewed as privileged because of the additional resources provided to set up and run laboratory activities. Unlike most other disciplines, science teaching requires dedicated space where materials can be set up. Historically, it has also been easier to seek external funding to support science and mathematics activities than those in some other disciplines, so science teachers are often perceived as having more advantages than other teachers. Our recommendations include: (a) when seeking external funding, seek project designs that are thematic and involve other disciplines; (b) show members of partnership how other disciplines such as math, social studies, English, art, or music drive science advancements and are used in science education; (c) be aware of vocabulary used that is specialized to science, and seek to build a common language among partners and their peer teachers; (d) develop and share student assignments that demonstrate that science is a way of thinking and is part of day-to-day life; (e) encourage other teachers and community members to be active in your classrooms; and (f) move beyond involvement in science-related activities, to take an active role in other projects.

MAINTAINING AND ADVANCING THE WORK

All partners must see rewards in the partnership, and make recognized contributions, or the partnership will fail as investment decreases. Given the evolving nature of this work, we see three main issues as critical to establishing and maintaining effective partnerships: responsiveness to changing institutional climates, quality assurance, and scholarship. These three issues are common challenges of school-university partnerships in which rewards and contributions must be considered and maintained.

Responsiveness to Changing Institutional Climates

Partnerships are formed to meet agreed-upon needs of each partner in particular situations. As the leadership and priorities of each partner institution change and evolve, partnerships must have structures or mechanisms in place to realign foci to meet emergent needs of partner institutions. To meet these needs partnerships engage in periodic evaluations, goal-setting cycles, and reformulations of staffing patterns. In some partnerships, these changes are tied directly to funding sources. Periodic renegotiation of roles, responsibilities and expectations is common as the partnership is established and grows. One unintended effect of partnerships can be the emergence of status hierarchies within partner institutions. For example, teachers involved in partnership activities may receive additional recognition or honors because of their work, or a teacher who has reassigned time to work on a project may be perceived to have a reduced workload. Within the ranks of a school faculty, a perception of 'haves' and 'have-nots' may emerge, leading to resistance to partnership activities, and schisms within faculties, departments, and teams. While some tension may be unavoidable, attention to professional relationships is a must for the longer-term vitality of the partnership.

Quality Assurance

Each partner institution may have well-established standards for activities and products. When activities and products of the partnership do not meet these standards, partners may 'feel the heat' from peers in their institutions, and the partnership may come under increasing scrutiny and disfavor in the absence of attention to these standards. Conflicting agendas and priorities are common, and require special attention to achieve some balance. For example, many states now have annual tests to measure student achievement or proficiency. Mentor teachers may perceive as risky to have a novice teacher take charge of a class prior to this test. (Novice teachers rarely support students to the level that experienced teachers can in preparing for these tests.) Likewise, universities cannot expect schools to replace curricula designed to help student prepare for these tests with pilot curricular materials developed by a novice. Thus, accommodations for school calendars and curricula must be made so that the partnership can thrive. Instead of refusing to place the novice teacher, a school might instead see an opportunity to support novice teachers as they learn how to help students prepare for these tests. (A coteaching arrangement such as that described by Roth in this volume would constitute a workable alternative to serve the needs of teacher and student teacher.)

Scholarship

The nature of university work means that most university partners must do some research as part of their work in a school-university partnership. Scholarship can also directly impact funding if the partnership relies on grants. The necessity for academic products is still high in most competitively funded projects. Both schools and universities can benefit from scholarly inquiry, and should be involved in production and dissemination. The *process* of research is commonly of more benefit to school partners than products themselves, in terms of improved teaching, learning, and curricula. Curricular materials, action-research reports, conference presentations, and formal research reports can be common products, as can project evaluations. For teacher participants, curriculum development activities, action research and formal studies of classroom interventions, teaching strategies, or learning situations may form the basis for professional growth. Most products can be used 'as is' or modified by university partners to demonstrate merit for promotion and tenure. These products can also be used to leverage additional funding for partnership work. (For another example of such leverage see the description of the Texas Regional Collaboratives by Barufaldi and Reinhartz, this volume.)

A partnership can take a great deal of time and energy to establish and maintain, and all partners must look for ways of incorporating different aspects of their work. At any given time in a school-university partnership, it is common to see three different kinds of questions being explored: action-research and teacher-research, university-partner–driven, and cross-institutional questions.

Action-research and teacher-researcher questions. Research is not merely the domain of university partners; school partners, too, have questions that they want to learn more about. Partnerships can provide extra support to enable teacher-directed research, and university partners can help school partners learn about and conduct action research. University partners and school partners working together can articulate questions for, design, and conduct classroom research. Research might examine a particular teaching strategy like the effectiveness of learning logs or journals. Research questions could also focus on the learning of a specific set of science concepts or theories. Other areas of research include gender issues (see Rennie, this volume) or at-risk students. Reflective teacher practice is a common programmatic goal for school-based activities of novice teachers. In many states, demonstrating a reflective stance toward teaching and learning is required for continued licensure, as measured by instruments like Pathwise or PRAXIS, and is a necessary prerequisite to National Board Certification. To promote reflection, pre-service teachers are often required to do some sort of classroom research, such as exploring student understanding of science topics or the effectiveness of particular teaching strategies.

University–partner-driven questions. University partners are often better positioned to think about broader questions related to teaching and learning, since they are not as closely tied to day-to-day classroom activities. Whether these questions are initiated by school administrators or university partners, it is common for university partners to take leadership roles in formulating questions and initiating research. Research questions can range from the effectiveness of programs or curricula, and how best to work with and support pre-service teachers, to the nature of scientific activity in schools and scientific communities, and the history, sociology, and philosophy of science. Whereas these questions may directly inform approaches to science teaching, they often focus more on the thought undergirding lessons than on the lessons themselves. By nature, they have less impact on day-to-day teaching than questions about teaching strategies and techniques.

Cross-institutional questions. Some kinds of research topics cut across partnership roles. These topics are found where interactions between partner institutions shape new working relationships and programs that each institution is interested in exploring. For instance, as partners work to understand how constructivist views of learning can inform teaching practice, they may choose to research, for example, issues of assessment and their relationship to practice, the role of mentoring and team approaches to instruction, or mentoring and collaboration models. The formation of these types of research foci can be a real indicator of the success of a school-university partnership. When partners from both institutions demonstrate ownership of the same sets of questions and share substantial roles in conducting and reporting research, success is indicated. In these situations, a strong school-university partnership may evolve—one that is strong enough to withstand invariably changing personnel and changing levels of resources over time.

EVALUATION AS A TOOL FOR SHAPING THE PARTNERSHIP

Evaluation of the school-university partnership can take a variety of forms. Whenever a goal is negotiated and defined, partners must also consider how they are going to evaluate their progress toward the goal as well as the appropriateness of the goal. While much of this chapter reflects this approach in explicit and implicit ways, we reiterate the most important principles here.

School–university partnerships are a form of long-term, evolving professional development for multiple audiences. Regardless of which partners take the lead in various activities, effective evaluation will provide important feedback to shape near-term activities, and benchmarks by which to measure the effectiveness of the partnership. The following evaluation questions are modifications of those used by professional developers (Loucks-Horsley, Hewson, Love, & Stiles 1998), tailored for the work of school-university partnerships.

What are the goals or desired outcomes of the partnership? Setting and negotiating goals is central to the success of a partnership. Being able to articulate these goals in ways that individuals outside the partnership understand them is key to obtaining wide-range external support needed to sustain and advance partnerships.

What are the most important outcomes to assess and why? Partnership work is varied, and includes new roles that may not be well understood by those inside or outside of the partnership. Therefore, it is essential that successes and challenges in these new roles be documented. Since it is likely not possible to evaluate all goals and projects initiated by the partnership, setting priorities will prevent the work of evaluation from becoming overwhelming.

How can these important outcomes best be measured? What information will be needed and who will take responsibility for collecting and analyzing the information? Collecting and analyzing information is a kind of work most often associated with school administrators and university partners. Whereas this may be appropriate in many settings, decisions about this must be made with the specific goal or project in mind. Some goals may call for all members of the partnership to assume roles in collecting and analyzing information. Setting a timeline with specific deadlines and expectations can enable this work.

How can evaluation contribute to continuous improvement and evolution of the partnership? Ongoing evaluation provides information needed to support, sustain and direct the work of a partnership. Evaluation sets the direction for the future. Members of a partnership must feel that their work is valued and has contributed to the success of the partnership. Evaluation provides mechanisms for documenting and valuing the work of many partners.

SUMMARY AND RECOMMENDATIONS

In this chapter, we examine partnerships from the perspectives of the planner, developer, and participant. We provide frameworks for thinking about school-university partnerships and the roles they can play in the preparation of science teachers. When a school-university partnership is formed, it initially reflects characteristics of both institutions, but it also has an identity of its own and characteristics that are unique to that partnership. If the partnership is to survive and flourish, these differences must be recognized and used to the benefit of the partnership. There is no single formula for developing a partnership, so selecting strategies and approaches that fit partnership goals is essential.

We believe that there is a set of characteristics shared by all successful school-university partnerships. A set of shared or common goals that have been identified and defined by the partnership is primary among these. Once goals are established, specific plans for addressing the goals should be developed. These plans must include timelines, address specific resource needs (time, individuals, materials, and funding), and include a clear sense of what the desired outcomes or products look like. Developing goals and related plans requires on-going attention to communication. Goals and plans should be open to modification as conditions change and as new members join the partnership. Strong documentation helps the partnership evolve and demonstrates how a partnership supports educational goals, even beyond those of its members.

We would not be honest about our understanding of school-university partnerships if we did not openly address the dilemmas of working in partnership settings. Differences in the cultures of schools and universities cannot be ignored. A growing body of literature highlights these differences, and the difficulties they can create in partnership relationships (Johnston, 1997). Our experience shows that when these differences are ignored, partnerships become weighty burdens for the participants, fail to evolve over time, and eventually dissolve. However, these differences should not be seen as a reason to avoid entering into partnership, nor should they diminish the realization of significant advantages that partnerships can bring to all stakeholders. School-university partnerships do provide unique settings for blending these cultures and for preparing science teachers to meet the challenges of teaching to today's standards, and those of the future.

NOTE

The vignettes in this chapter are based on partnership relationships between Michigan State University and its Professional Development Schools, and in particular the authors' work in the Holt Public School District. The authors are deeply indebted to the staff of Holt Public Schools for teaching us how to be partners as we represented the University in building relationships with schools. These vignettes are also shaped by our work in Charlotte (North Carolina), Columbus (Ohio), and Toledo (Ohio) area schools.

REFERENCES

American Association for the Advancement of Science. (1998). *Blueprints for reform.* New York: Oxford University Press.

Coalition of Essential Schools (1998). http: /www.essentialschools.org/.

Cushman, K. E. (1993). Teacher education in Essential Schools: The university-school partnership. *Horace, 10*(1), 1-10.

Day, C. (1998). Re-thinking school-university partnerships: A Swedish case study. *Teaching and Teacher Education, 14*, 807-819.

Johnston, M. (1997). *Contradictions in collaboration: New thinking on school/university partnerships.* New York: Teachers College Press.

Levine, M., & Trachtman, R. (Eds.). (1997). *Making professional development schools work: Politics, practice, and policy.* New York: Teachers College Press.

Loucks-Horsley, S., Hewson, P., Love, N., & Stiles, K. (1998) *Designing professional development for teachers of science and Mathematics.* The National Institute for Science Education. Thousand Oaks, CA: Corwin Press.

National Network for Educational Renewal. (1999). *Partner school compact.* http://weber.u.washington.edu/~cedren/PartnerSchoolCompact.html.

National Research Council. (1996). *National science education standards.* Washington, D.C.: National Academy Press.

Sobel, D., French, N., & Filbin, J. (1998). A partnership to promote teacher preparation for inclusive, urban schools: Four voices. *Teaching and Teacher Education. 14*, 793-806.

The Holmes Group. (1986). *Tomorrow's teachers.* East Lansing, MI: The Holmes Group.

The Holmes Group. (1990). *Tomorrow's schools.* East Lansing, MI: The Holmes Group.

The Holmes Group. (1995). *Tomorrow's schools of education.* East Lansing, MI: The Holmes Group.

Wasley, P. (1991). *Teachers who lead: The rhetoric of reform and the realities of practice.* New York: Teachers College Press.

THE DYNAMICS OF COLLABORATION IN A STATE-WIDE PROFESSIONAL DEVELOPMENT PROGRAM FOR SCIENCE TEACHERS

James P. Barufaldi & Judy Reinhartz
University of Texas

INTRODUCTION

The historical roots of professional development are steeped in attitudes that portray teachers as semi-skilled and teaching conditions that can be described as anti-professional and full of uncertainties (Darling-Hammond, 1994). Such attitudes and working conditions in teaching have helped shape professional development programs, often referred to as staff development or inservice. Teachers are brought together and shown how to use curriculum, commonly characterized as teacher-proof because "even teachers cannot ruin it". Teachers then, are only implementors who can be "trained" to use the curriculum and materials in their classrooms (see also the chapter by Roth). The maxim "Those that can't, teach" has long been used to characterize teaching as a bleak profession with few opportunities for teachers to be leaders.

To adequately review the professional development programs of the last three decades necessitates placing them in a historical context. These programs used a show-and-tell approach that required teachers to be passive as they sat and listened to experts, who would come and spend a day or less to tell them how to do their job. The interaction with the audience was limited, yet teachers in attendance were expected to implement in their classrooms what they just "learned" (Loucks-Horsley, Stiles, & Hewson, 1996). This expectation becomes problematic because extreme isolation from other teachers and adults—often limited to 20–30 minutes per day during lunch (Wasley, 1991). Professional development programs are "everything that a learning environment shouldn't be: radically underresourced, brief, not sustained, [and] designed for 'one size fits all'..." (Miles, 1995, p. vii).

These views of professional development programs, according to many, lack intellectual coherence and certainly lack an appreciation for teachers and the teaching profession. Historically, teachers have participated in sessions that operated out of a deficit-model; the basic premise is that something is wrong, and needs to be fixed. In addition, teachers are not considered creators of these sessions, merely the recipients of the training provided. It is this perception of teachers and teaching that has guided professional development over the past several decades.

Professional development programs could be places of learning for teachers (Hargreaves, 1995). This would require perceptions of teachers as (a) leaders who

D.R. Lavoie and W.-M. Roth (eds.), Models of Science Teacher Preparation, 89–105.

want to learn, (b) constructors of new knowledge in an environment that is *collaborative* and safe for experimentation, and (c) able translators of these experiences into developmentally appropriate curriculum materials. If these three assumptions are subscribed to, then professional development programs can be classified as competency-based (Smylie & Conyers, 1991). A competency-based approach to professional development empowers teachers (Gilbert, 1994) and communicates to them that their knowledge, skills, and experiences are considered assets. In such an approach, teachers are valued for what they know about science, what they know about effective teaching, and what they know about how students grow and learn.

The theoretical base supporting such assumptions and perceptions of teachers and teaching comes from two different theories—constructivist theory and sociocultural theory. Constructivist theory has Piagetian roots; sociocultural theory has Vygotskian roots (Hatano, 1993). Taken together, these theories drive a professional development model for science teachers—one that is sensitive to teachers and who they are, how they learn, and what role theory plays in the teaching and learning process (Howe & Stubbs, 1996). Constructivist theory recognizes teachers as active agents in the construction of their knowledge. Such a view was largely ignored until 1978 when Driver and Easley (1978) provided the vocabulary for describing the learning process of teachers.

The sociocultural theory legitimizes teachers' work with colleagues in ways that provide a forum for problem solving and dialogue (see Roth, this volume). In such dialogue, the language serves to mediate ideas as teachers work their way through scientific misunderstanding and misconceptions. In a competency-based professional model, teachers are empowered and independently come up with their own reasonable explanations derived from dialoguing and opportunities for personal and professional reflection. In this rich environment, teachers do not expect others to provide the answers to their questions or to resolve their problems; what they expect is to have a sounding board for testing ideas and possible solutions.

The theoretical base, including such assumptions and perceptions of teachers and the profession supports the development of teacher leadership skills. The Texas Regional Collaboratives for Excellence in Science Teaching, a statewide professional development program for science teachers, is one that holds promise and can serve as a model.

In this chapter, we discuss the Texas Regional Collaboratives for Excellence in Science Teaching Model and the changing roles of various educational entities including the public schools, businesses, communities, universities, and Education Service Centers in delivering a quality science professional development program. The cornerstone of the model is collaboration, "together we can make a difference". (Fetters and Vellom, this volume, present a different collaborative model.) It is the bringing "together" that epitomizes the model of collaboration. Teachers thinking in isolation gives way to collective thinking and sharing; they are not alone in their pursuit of science teaching excellence. Members of the collaborative are important links in the chain for providing quality professional development. The Model is "everything that a learning environment [should] be": long term, sustained, supported, and designed to meet the needs of practicing science teachers (Miles, 1995, p. vii). Science teachers who want to learn are capable of constructing new knowledge and they translate what has been learned and practiced into developmentally appropriate experiences for their students.

The professional development model operates within a collaboratively driven organizational system built on a shared vision and collegial talk. The model of collaboration achieves the goals that Lieberman (1996) discusses because teachers who have opportunities dignify their experiences through sharing which may lead to fundamental changes in their classrooms. Sharing a vision and talking with colleagues also have the benefit of bringing participants together. As relationships grow and deepen, the teacher network strengthens and becomes a catalyst for change. In each collaborative, this social context of authentic partnership exists that makes the story of the model of collaboration so poignant in promoting, not only professional development, but personal growth.

SETTING THE STAGE FOR COLLABORATION

Since the publication of *A Nation at Risk* (Gardner, 1983), educators, community leaders, representatives from the private sector, and parents expressed great concern about the poor preparation of students and their respective performance in science and mathematics and called for major reform of education. The theme of the publication focused on economics as a basis for promoting the study of science and technology in American society. Many concluded that the United States was loosing its economic competitive edge in the global market place. The late 1980s found science education in yet another new age of reform (Bybee, 1993). During the 1980s and 90s more than 500 national reports addressed various inadequacies in science curricula and in the preparation of teachers. Concerns, centered on inservice education of science teachers, are reflected in recent school reform initiatives that have been designed, funded, and implemented in the United States. Projects such as the statewide rural, urban and local systemic initiatives, and the *National Science Education Standards* (NRC, 1996), mostly funded by the National Science Foundation (NSF), the U. S. Department of Education, and many private foundations, businesses, and professional organizations, are impacting elementary and secondary education. These projects and initiatives have become catalysts in providing exemplary staff development opportunities for teachers. These reform efforts focused on a key construct, *systemic*. Systemic change, the force behind school reform efforts, has emerged as an educational movement. Systemic school reform embodies three integral components: (1) the promotion of ambitious student outcomes for all students; (2) alignment of policy approaches and the actions of various policy institutions to promote such outcomes; and (3) restructuring of the governance system to support improved achievement (Goertz, Floden, & O'Day, 1996).

Currently, NSF has committed approximately $283 million dollars to systemic reform initiative efforts. NSF initiatives such as those mentioned above, endorse the following critical developments that are fueling systemic change efforts (Division of Educational System Reform, 1997, p. 2) and include the following:

1. Implementation of comprehensive, standards-based curricula as represented in instructional practice, including student assessment, laboratory, and other learning experience provided through the system and its partners.

2. Development of a coherent and consistent set of policies that supports provision of high quality mathematics and science education for each student; excellent preparation and continuing education, and support for mathematics and science teachers (including all elementary teachers); and administrative support for all persons who work to dramatically improve achievement among all students served by the system.
3. Convergence of the usage of all resources that are designed for or that reasonably could be used to support science and mathematics education—in fiscal, intellectual, material, curricular, and extra-curricular ways—into a focused and unitary program to constantly upgrade, renew, and improve the educational program in mathematics and science for all students.
4. Broad-based support from parents, policy makers, institutions of higher education, business and industry, foundations, and other segments of the community for the goals and collective value of the program. There should be rich presentations of ideas behind the program, the evidence gathered about its success and its failures, and critical discussions of its efforts.
5. Accumulation of a broad and deep array of evidence that a program is enhancing student achievement, constituted through indices that might include achievement test scores, higher level courses passed, College admission rates, college majors, Advanced Placement Tests taken, portfolio assessment, and ratings from summer employers. There should be evidence that students generally achieve at a significantly higher level than previously.
6. Improvement in the achievement of all students, including those historically underserved.

Systemic school reform calls for a new way of thinking about change and innovative strategies to initiate such change. Independent strategies are not appropriate given the enormity and complexity of school reform efforts. A systems perspective on school reform directs attention to the necessity to pursue educational change through a strategy of *active collaboration* among all major stakeholders or partners in the system. This strategy assumes that more can be achieved or accomplished jointly than individually through collaboration and has major implications for science teacher professional development.

The success of a collaborative effort is related to certain characteristics of its members (Mattessich & Monsey, 1992). These authors note that characteristics consist of "skills, attitudes, and opinions of the individuals in a collaborative group, as well as the culture and capacity of the organizations which form collaborative groups" (p. 19). Individuals functioning in collaborative groups share an understanding and respect for each other and their respective organizations in the way they operate as well as for cultural norms and values, limitations, and expectations of the organization. Collaborating partners are able to compromise, are flexible, open to suggestions, and acknowledge that conflict is good and it is acceptable to agree or disagree.

Unfortunately, many believe that the process of collaboration does not require any special skill among members and that collaboration occurs spontaneously when more than one individual come together to pursue a common goal. Collaboration, however, is a *learned process*, one that must be practiced, nurtured, and supported.

NATURE AND PROCESS OF COLLABORATION

Collaboration is the cornerstone of developing and implementing successful science professional development. It is also a major theme that permeates the National Science Education Standards (NRC, 1996) as evident in the following statement: "Science learning experiences for teachers must encourage and support teachers in efforts to collaborate" (p. 4). The Texas Regional Collaboratives for Excellence in Science Teaching Model is used to illustrate various components of the collaborative process. What is the organizational system of the Regional Collaboratives?

The administrative office for the Texas Regional Collaboratives resides in the College of Education's Science Education Center at the University of Texas at Austin. Briefly, the major goal of the 20 Texas Regional Collaboratives is to create ongoing partnerships of educators and business leaders who are committed to science education school reform. Regional Collaboratives are partnerships among local colleges and universities, education service centers, school districts, business and industry, informal education sites such as the Texas State Aquarium and the Centennial Museum, and the community.

A collaborative organization must have an adequate and ongoing financial support base for its many activities. Obtaining such support is necessary for continued implementation of programs; therefore, partners work together through cost-sharing, in-kind contributions, and pooling of human resources to develop and sustain high quality professional development opportunities for K–12 science teachers. Major financial resources of this project were provided by NSF, the Texas Education Agency/U. S. Department of Education Dwight D. Eisenhower Program, businesses (e.g., Casio, Southwestern Bell Communications Foundation, Exxon, Shell, Holt, Rinehart, and Winston, Delta Education, and Apple), and others. The ultimate goal is to improve the teaching and learning of science and to engage all students in interesting, relevant, experiential, and meaningful science learning experiences (Jbeily & Barufaldi, 1998).

Twenty regional sites, staffed by instructional teams of professors from local universities, community college instructors, science supervisors and coordinators, and master teachers from local high schools provide science instruction, approximately 100 contact hours per academic year, to practicing science teachers in their region. Each collaborative is autonomous. Yet they subscribe to important common elements of professional development. These include (1) a commitment to collaboration, high standards, alternative assessment, experiential learning, and constructivism; (2) a philosophy of bringing the real world into the classroom; and (3) an integration of instructional and communication technology into their educational programs (Jbeily & Barufaldi, 1998). The common elements are consistent with those presented in the National Science Education Standards (NRC, 1996) in that more emphasis is on the teacher as a member of a collegial professional community and that teachers are involved in collegial and collaborative learning opportunities.

Barufaldi and Reinhartz (1998) identified four components—shared vision, interconnectivity among the organizational units of the system, recognition of a multi-tiered process, and support within the system—that are imperative to collaboration. The boundaries between these units are generally seamless in that the initial shared vision may evolve and change as the result of various social, economic, and political pressures that operate within and outside of the system, as they have done in the past (Barufaldi, 1977). Changes in goals and objectives, funding patterns, support, human resources, personnel, and state and federal mandates, may give rise to the rethinking of purpose and mission, which may eventually result in a newly created vision within the system.

Shared Vision

The most important component in the collaborative process is a *shared vision.* This vision evolves from goals and objectives that embrace good practice and reflects a degree of commitment among the partners. "Collaborating" partners have the same vision, with clearly agreed upon mission, objectives, and strategy (Mattessich & Monsey, 1992). The shared vision may exist at the outset of collaboration, or the partners may develop a vision as they work together. The authors believe that collaboration is a mutually beneficial and well-defined relationship entered into by two or more organizations to achieve common goals. When collaboration is successful, there is commitment to (1) a definition of mutual relationships and goals, (2) a jointly developed structure and shared responsibility, (3) mutual authority and accountability for success, and (4) shared resources and rewards (p. 40).

Interconnectivity

In order for collaboration to be successful, this shared vision must be communicated throughout a complex organizational system such as found in higher education. The *interconnectivity* of this vision is essential among the organizational components of the system. The shared vision assumes a greater degree of specificity, top down, as it is operationalized throughout the system—university, colleges, departments, centers, and programs such as the Texas Regional Collaboratives; yet, individuals with responsibilities in the organization must embrace and support the "essence" of the vision. A clear understanding of the *interconnectivity* among and between individuals within the organizational units is necessary for the system to function and be productive (Barufaldi, 1998). Recognition of the interconnectivity of the structure as to how the vision is translated at each level helps answer the question, "What makes collaboration work in science teacher professional development?"

Members within the collaborative must also sense a degree of ownership and commitment. Working within a system makes one aware that interconnectivity also has a human side to it. Thus, "The task for educational reform in the sciences is generally seen as one of developing a vision that recognizes that science, technology, society, and the quality of human existence are interconnected, and that their borders are disappearing" (Hurd, 1997, p. 9).

Ownership is important, which is reflected in the *Educating Americans for the 21st Century*. Thus, "the educational system must provide opportunity and high standards of excellence for all students—wherever they live, whatever their race, gender, or economic condition, whatever their immigration status or whatever language is spoken at home by their parents, and whatever their career goals" (NSB, 1983, p. 8). The system must be sensitive to the needs of all; the challenge is to ensure that our shared vision is truly part of our personal agenda. A personal agenda implies that we must instill in our profession the notion of *ownership* and encourage the feeling of ownership among students, parents, and the community during active collaboration. *Ownership reflects a special relationship, the interconnectivity that forms a bond between individuals, groups, and systems* (Barufaldi, 1987).

A feeling of ownership is enhanced when the following dimensions are realized (Mattessich & Monsey, 1992). First, there is adequate time and resources must be devoted to developing the collaborative effort. Second, the operating principles and procedures of a collaborative group must promote among members a feeling of ownership about decisions and outcomes. Third, the ownership of a collaborative group should be continuously monitored over time and changes made to ensure the feeling of ownership. Fourth, interagency work groups should participate in regular planning and monitoring of the collaborative effort and thereby solidify ownership and ongoing commitment. The interconnectivity is dynamic within all organizational units within the system, viewed as part of a human endeavor, and forms long-lasting bonds of ownership and commitment among members of the collaborative.

A Multi-Tiered Process

The degree or intensity of collaboration within a system is part of a *multi-tiered* process. For example, the shared vision of the Texas Regional Collaboratives is to improve the teaching and learning of science by providing teachers of Texas with opportunities to participate in a program of sufficient intensity and duration to have a positive impact on their performance and student achievement. Differences lie in teachers' experiences, the nature of the experiences, and the conditions under which they experience the program. As the collaborative process evolves, major stakeholders may change or experience a shift in their roles, the social context may change, and the initial shared vision modified as it becomes operationalized through a particular infrastructure, which eventually delivers the "essence" of contemporary science education.

Within higher education, all partners or stakeholders must demonstrate support and be fully committed to the vision. Within a complex system, collaboration is viewed as a multi-tiered process with many nodes. At each level of the system the degree of collaboration varies as the shared vision evolves over time. For example, the University of Texas at Austin is committed to serving the people of Texas by supporting and nurturing programs that address their needs and concerns. Diversity, inclusion, and serving underprivileged populations through many special projects and outreach programs play a major role in the University's mission. (Regarding these special issues, see also the chapters by Lee/Fradd and Thompson *et al.*) Collaborative activities at the university level result in a vision that reflects a more general notion of improving the teaching and the learning process among all students, levels and disciplines.

At the college of education and the department level, more individuals have the opportunity to play an active role in the collaborative process as the general notion of shared vision becomes more delineated through the major activities found within academia—service, teaching, and research. It must be pointed out that the vision of improving teaching and the learning for all is paramount in all professional activities conducted at the College and Department levels.

When one views this shared vision in a specific discipline or center such as the Science Education Center at the University, the vision becomes more focused as it specifically addresses the improvement of science teaching and learning through the development and the implementation of a teacher professional development program. The Texas Regional Collaboratives operationalizes the program. Many professional educators are involved at this level as they plan, design, implement programs, and evaluate outcomes of newly created opportunities for teachers.

At the level of the Texas Regional Collaboratives, collaboration is more intense and involves many more individuals. These include master teachers, instructors, business partners, science and mathematics educators, scientists, mathematicians, and administrators. The vision of improving the teaching and the learning of science also acknowledges the social context, and the unique political, economic and social factors within the particular region of the state. The vision is personalized as it channels and "molds" the thinking and learning processes of both teachers and students. This shared vision becomes the synergistic component that propels the Texas Regional Collaboratives to fulfill the commitment to professional development. The process of collaboration helps build a community of learners that continues to support and nurture a shared vision. We anticipate that many partners in this collaborative process become transformational leaders who will improve the teaching and learning of science through innovative professional development programs.

Support

The need for support throughout the system was briefly alluded to above. We stress that an adequate financial base to support the operation and activities of the collaborative is of highest priority in initially forming and then sustaining the group. Funds must be generated in support of such efforts. Collective grant writing and fund raising are crucial. Furthermore, solicitation of support from business and industry is best determined by asking the question, "What is in it for the potential funder?" (Jbeily & Barufaldi, 1998)

Successful collaboratives use federal, state, and local monies as leverage to attract additional funding. Collaboratives with a relatively long history subscribe to the solicitation of funding as a collaborative process; resources are pooled or jointly secured. Many government initiatives and (private and public) foundations require the pooling of resources and tend to discourage requests for funding from entities that wish to work autonomously.

Support also includes incentives and rewards that are part of the system. For example, the incentives for practicing teachers involved in professional development programs include (a) release class time to participate in such programs, (b) special recognition from the school district, region, and state, (c) special stipends, (d) provision of child care if necessary, (e) opportunities to work toward an advanced degree which includes free tuition and reimbursement for books and related materials, (f) granting of monies to support classroom activities, and (g) appropriate funding to attend conferences sponsored by various science teacher organizations. Support for other educators involved in professional development programs may include (a) a reduced "teaching load" necessary to conduct activities of the program, (b) additional graduate assistants to assume some of the responsibilities, (c) a clinician to assist with classroom observations, and (d) a reward system that acknowledges one's involvement in this type of labor intensive professional endeavor.

ATTRIBUTES OF THE REGIONAL COLLABORATIVES

What are the attributes of the Texas-Regional-Collaborative professional-development program? There are three salient attributes that provide the infrastructure including instructional teams, partnerships, and fund leverage. Through active collaboration, these attributes have helped to institutionalize the program regardless of geographic location in the state, the number of science teachers involved and their grade levels.

Instructional Teams

The first attribute is the establishment of *instructional teams*. The composition of the teams include such individuals as scientists from the university and industry, master mentor teachers from local school districts, instructional specialists from educational service centers, or community members from museums, zoos, or nature centers. The members of the instructional team are varied and diverse with regard to their positions, areas of (personal and professional) interest, and levels of expertise.

According to the National Science Education Standards (NRC, 1996), "The challenge of professional development for teachers in science is to create optimal collaborative learning situations in which the best sources of expertise are linked with the experiences and current needs of the teachers" (p. 58). This standard is operationalized as each instructional team provides more than 100 contact hours of professional development science education to 25 teachers. This development is designed to engage teachers in learning experiences that are well aligned with changes in curriculum, assessment, instructional strategies, and advances in science and technology. During the academic school year, each teacher is provided incentives such as stipends, materials, and release time from their regular duties, to mentor five teachers in their district.

It was determined quite early in the design of the Texas Regional Collaborative program that instructional team members also needed opportunities to grow professionally; therefore, a series of professional development academies were designed. Four to six workshops are offered each year to members of each instructional team. The workshops are aligned with the goals of the program and focus on topics such as instructional technology, constructivism, implementation of nationally recognized science programs, alternative assessment, reading in the content area of science, and the national and new state standards.

Shared Systemic Threads

All instructional team members model and use (a) a variety of instructional strategies, (b) educational technology, (c) alternative assessment approaches, and (d) state mandates (Texas Essential Knowledge and Skills) which emphasize the national standards. These four systemic threads form the common denominator for each of the 20 instructional teams.

The first thread, instructional strategies, embodies a constructivist-sociocultural theoretical base. This emergent sociocultural perspective recognizes that "learning is a constructive process that occurs while participating and contributing to the practices of local community" (Cobb & Yackel, 1996, p. 195; see also the chapters by Roth and Thompson and Hargrave). As teachers work in cooperative-learning groups, they have opportunities to interact and network. They expand on what they already know by constructing new knowledge or by reinterpreting what they already know against what they are currently learning. When teachers achieve new constructs through assimilation and accommodation the process is self-regulated. Learning that is meaningful does not come as a result of accumulation, but rather it results from reflection and resolution to a series of cognitive conflicts in a safe, nonthreatening environment of the Texas Regional Collaboratives.

Educational technology for teaching and communicating, the second systemic thread, connects the knowledge learned to a wider audience in a more effective and efficient way. Teachers in the Texas Regional Collaboratives experience technology as they conduct experiments and investigations in the field or in the classroom. (See the chapter by Lavoie in this volume for a concrete example of technology use in science teacher preparation, involving many of the following topics.) They learn to use graphing calculators that record data from probes for testing temperature, light, sound, pH of solutions, and heart beats of animals and humans. Once these data are recorded and entered into the calculator or computer, they are presented in graphic form; this information could then be analyzed and interpreted. Accessing information on specific science topics under investigation from the Internet was another way educational technology was implemented in science classrooms. Technology as a research tool provided teachers with opportunities to increase their resources by printing out a hard copy of helpful information. The digital camera became an integral part of teachers' instructional repertoire by recording events and activities as they occurred in the classroom. These digital photographs can be downloaded into the computer. Hard copies could be printed which some teachers included in their professional portfolio, or these photos could be incorporated into electronic slide programs for parents' night and school board meetings. In addition, these photos could be used for instructional purposes or as evidence of professional growth for individual teachers as well as students. Lastly, teachers in cooperative-learning groups prepared a computer presentation on an agreed-upon science topic that was presented to other teachers. By developing a computer presentation using a specific software program in small groups, teachers have opportunities to take what they have learned and create their own classroom presentations. All these experiences helped teachers become more proficient in using educational technology in their classrooms.

The third thread concerns alternative assessment. Teachers in each collaborative are encouraged to use alternative forms of assessment in their classes once benchmarks have been established with ample opportunities for practice. (For a concrete case of using portfolios to integrate theory and best practice see the chapter by Stein in this volume.) As teachers study and learn about new topics, they consider using performance assessment and develop rubrics for measuring achievement of science lessons and unit objectives. The development of these performance measures and rubrics is collectively accomplished as teachers share and produce a joint product. Since teachers learn in different ways, there should be different approaches used to measure their learning as well. Embedded in the overall assessment procedures used in the Texas Regional Collaborative model is a series of measures which included, but were not limited to, reflection journals, portfolios, and performance tasks.

The last systemic thread that all Regional Collaboratives share is implementation of the state mandates for science along with those recommended at the national level. In Texas, the state-mandated *Texas Essential Knowledge and Skills* will be the engine for science curriculum and instruction during the next several years. The *Texas Essential Knowledge and Skills* are compatible with the national standards and form the backbone needed to teach science in a way that embodies constructivist, sociocultural, and hands-on/minds-on perspectives.

Partnerships

Another salient attribute shared by all Texas Regional Collaboratives is the support from business partners. These business partners not only provide expertise, but often materials, equipment, and money for field trips, supplies and dinners not available from grant funding sources. Industrial scientists are often on the cutting edge of research and as members of the instructional team, they become stakeholders in the goals and objectives of the Texas Regional Collaboratives. The support provided by these partners enriches the type and quality of the professional development experiences that can be and are provided to science teachers in the individual collaboratives. Business partnerships position the collaboratives positively in the community in the pursuit of enhancing the science learning of both teachers and their students.

These business partners are often locally based to better support the Texas Regional Collaborative in that area. Visits to these businesses help teachers to better understand where and under what circumstances research is conducted, types of equipment required, steps taken before a product is marketed, and ways to protect the environment when disposing of wastes they generate. There are business partners who contribute at the state level as well, so that all the regional collaboratives can benefit. For example, by partnering with Casio, purchasing equipment packages, including a digital camera, science probes, and a graphing calculator, was made possible at an attractive price. Delta Education, another partner, has provided materials, kits and necessary training for specific collaboratives enable many teachers to experience the activities and to use them in their own classrooms. EXXON, Southwestern Bell, Shell Oil, and Apple Computers support the Collaboratives in a number of ways. They have provided funds to underwrite such activities as 'Honoring the Teachers' luncheons, funded travel to state and meetings including the Conference for the Advancement of Science Teaching and the National Science Teachers Association, and purchased additional equipment for teachers to use. The professional activities made possible by these funds help to promote leadership capacity for those science teachers who participated. These teachers become leaders in their classrooms, at their school campuses, and in their school districts when they share what they have learned with their students, colleagues, parents, and policy makers. Taken together, there are more than 50 partners who support a variety of the activities of the Regional Collaboratives in the state.

Funding Leverage

All Regional Collaboratives share the last attribute, *funding leverage*. Business partners have made it possible for three million dollars of state-provided funds to be leveraged to more than eleven million dollars in total funding for the Collaboratives. It is clear that the Texas Regional Collaboratives accomplished much with fewer dollars from federal and state grant funding agencies. In some cases, in-kind funds are provided by scientists at the university and in industry who give of their time to make presentations and by community and educational agencies that provide space, expertise of staff, and waive entrance fees to museums and local amusement parks.

Funds are also leveraged through cost sharing or matching funds. Some businesses are willing to provide actual dollars to support work study or graduate students. Such funding sources are often modest but form the nucleus for adding funds from federal and state grants. The leveraging of funds through cost sharing and in-kind contributions by human and financial resources has enabled the collaboratives to build the necessary infrastructure to sustain a state-wide professional development program that is of sufficient intensity and duration to have positive impact on teacher performance and student achievement.

Value Added Dimension

While these salient qualities are extremely important, they are not sufficient to keep the Texas Regional Collaboratives program functioning effectively. There needs to be a value-added dimension. What then holds these three attributes together, helping them function in concert with one another effectively and efficiently? A shared vision, collaborative process, and the development of leadership capacity, and collaborative products form the *connective tissue* which hold the three attributes of instructional teams, partnerships, and leveraging of funds together.

Teachers in the collaboratives are more confident, enthusiastic, and knowledgeable; they talk in specifics. For example, they share ways to adapt materials to meet the needs of individual students. They are risk takers, have a better sense of what needs to be done, and do not wait to be asked to carry out a task. A second year teacher stated that

> this program has proven to be a big benefit as far as understanding science concepts and how to teach them to students. I have been challenged to broaden my teaching skills, and I can see the benefits in my classroom. I have met with my cadre a few times and I feel this is a benefit to my school because except for the classes I have been leading, there really has not been training available for our science teachers. My principal is extremely excited because she can see what I am able to bring to the school from the program.

Following Darling-Hammond's advice, the collaboratives are putting "greater knowledge directly in the hands of teachers... that focus attention on 'doing the right things' rather than 'doing things right' (1996, p. 7). According to principals, these teachers need less of everything; they are more skillful at managing students, at engaging students and keeping their interest because they have a repertoire of strategies. They are more child-driven and have a more comprehensive view of planning for science instruction. They are better at self-analysis and self-reflection because they are more aware of what it takes to be effective, and most importantly, are more honest with themselves.

Table 1. Benefits to Stakeholders

Stakeholder	*Benefits*
Schools	1. Better prepared science teachers. 2. Increased student achievement. 3. Reached all students using a variety of strategies - hands-on, technology.
Teachers	1. Empowered teachers to make needed changes in their science instruction, curriculum, and assessment including alignment. 2. Increased self esteem. 3. Assumed additional leadership roles. 4. Expanded knowledge - science content, use of technology, alternative assessment as well as the delivery system used. 5. Enhanced professional growth by attendance state, regional, and national conferences.
University	1. Focused on commitment to systemic reform. 2. Shared resources - professor expertise, facilities, and materials. 3. Increased graduate enrollment. 4. Built university collaboration. 5. Improved university teaching using diverse teaching strategies.
Public and Business Community	1. Used effective means to solicit federal/state/partner funds. 2. Increased quality of science teaching. 3. Improved education of work force. 4. Built connections to the real world of science and technology. 5. Partnered in promoting science education. 6. Aligned public and state sector with reform initiatives.

A "good" teacher education program has a "clear direction of improved practice, incentives for change, and ready access to help in planning and implementing change" (Andrews, 1997, p. 174). The regional collaboratives have clear visions for improving science education, provide teachers with incentives to make these changes in the form of equipment, resources, and opportunities for practice, and teachers have access to information and resources to implement these changes.

The Texas Regional Collaboratives develop *leadership capacity* at a variety of levels and in a variety of ways. The role of leadership means implementing "such mundane activities as making phone calls, raising money, arranging meetings, brokering resources and people, and negotiating time commitments for university and school-based educators" (Lieberman, 1996, p. 53). Teachers involved in this statewide program have demonstrated leadership by mentoring others, providing peer support, coaching on their school campus or with the other collaboratives via e-mail and telephone. The teachers have conducted workshops, presented at regional, state, and national meetings, and conducted action research projects in their respective classrooms. They lead discussions at meetings of the collaboratives and serve on advisory boards. The teachers also produced many *collaborative products* as the result of their participation in the program. A teacher directory of members of the collaborative for easy reference to information was produced. Some teachers designed and developed a newsletter, generated a historical record of events through pictures using digital cameras, collected and distributed research articles and resource information to keep members current, and organized a parental involvement science night. Others wrote mini-grants to support instructional and communication technology in classrooms. Here, not regulations but teachers transform schools as they collaborate with parents and administrators. It has been documented (Jbeily, 1998) that the Regional Collaboratives brought major benefits to major stakeholders (Table 1). Our experience is a concrete example of a community that grows and deepens, and thereby embodies a network that becomes an important catalyst for school renewal (Lieberman, 1996). It readily becomes apparent that meaningful collaboration resulted in a long lasting infrastructure that emerged as a network and delivery system for "best practice."

LESSONS LEARNED

Reflecting on the process of building the Regional Collaborative organizational system of human and financial resources to support and sustain quality opportunities produced the following.

1. Collaboration, working together, rather than alone, can produce very beneficial results.
2. Collaboration is a learned process.
3. Incentives and rewards are imperative to sustain meaningful collaboration.
4. The degree or intensity of collaboration within a system is a multi-tiered process.
5. The interconnectivity of the components of the system must be well articulated.
6. Support and encouragement are necessary to create linkages among the partners, and to help them realize that all can learn from others.

7. All partners will not contribute equally; recognize the strengths and weaknesses of each partner; remember that all members should have equal opportunity to contribute.
8. Meaningful collaboration takes time, perseverance, and patience.
9. The collaborative unit must have an adequate, consistent financial base to support its activities.
10. It is important to focus on the shared vision. The vision should be the "guiding light" to direct and modify your course of action, if necessary.
11. A collaborative relationship requires short term and long term planning and well defined channels of communication that support all levels of the collaborative process.

ASSESSMENT

The documentation of positive outcomes of the Texas Regional Collaboratives for Excellence in Science Teaching is aligned with the mission. The mission of the program is to provide professional development for science teachers that is ongoing and of high quality and intensity. This model program has been in operation for nine years and continues to bring colleges, universities, businesses, education service centers, and school districts together and work collaboratively. The sustained nature of the program has empowered more than 7,000 teachers to implement effective science practice in their classrooms by providing them content knowledge, training in correlating local, state, and national standards to curriculum, and pedagogy that will improve their performance and the science performance of their students. More than 600,000 students across Texas have benefited from improved instruction and exemplary instruction of teacher participants.

Another positive outcome is leveraging funds. During 1996-97, the support system has enabled the Collaboratives to leverage more than $1,619,957 of cost sharing/in-kind contributions for school districts Eisenhower formula funds, colleges and universities, and resources of local businesses and community partners in support of science teachers.

The program resulted in improvement in teacher performance and classroom instruction as reported by teacher observations, student comments, and feedback from administrators. The program continues to have a positive impact on teachers' knowledge and skills as evidenced by improved scores on post-tests. It appears that meaningful collaboration has resulted in an infrastructure that provides a delivery system of resources needed to implement a comprehensive, standards-based science curriculum.

One question that continues to challenge science educators is "Does the professional development model promote and enhance student learning?" The Collaboratives are addressing this question by conducting case studies to ascertain the degree to which teachers' professional development impacts student achievement. In Texas, student achievement in science is assessed the first time in grade 8. While measures of student achievement are showing positive results based on conversations with teachers and feedback from other participants, minimal formal test measures have been used statewide to determine the positive impact of the professional development model has had on student achievement. An important goal for the Texas Regional Collaboratives in the 21st century is to ensure (based on test results and portfolio indicators) that the gain in teachers' knowledge, skills, and performance has a positive impact on student achievement.

The most important lesson to be shared is that the "collaborative spirit" can affect school reform at every level but most importantly from inside the classroom. Teachers working with teachers, generate their own knowledge that is relevant for them. They are no longer consumers of expert knowledge, but producers. Such production was made possible through the ethos of the collaborative spirit. The essence of this spirit is evident in the following statement:

> Collaboration among the people involved in programs, including teachers, teacher educators, teacher unions, scientists, administrators, policy makers, members of professional and scientific organizations, parents, and business people, with clear respect for the perspectives and expertise of each. (National Science Education Standards, 1996, p. 70)

In the final analysis, the quality of teaching does not solely depend on the nature of the individual who enters and stays in the profession; it depends on the quality of the professional development experiences. Successful models of professional development have teachers work in collaboration with others and thereby transform science education. The model of the Texas Regional Collaboratives meets these criteria. It brings together teachers in an academic setting that is inquiry driven and practice-oriented in the pursuit of a common goal, which is "excellence in science teaching." In such a setting, teachers strive to better understand science content and ways to teach science. The model of collaboration presented in this chapter is a part of the "quiet revolution" for reconfiguring professional development because teachers are bringing about change in their classrooms (Darling-Hammond, 1996). As science educators look toward the 21st century, the goal will be to bring theory and practice closer together, and the Texas Regional Collaboratives for Excellence in Science Teaching model is one way to achieve it.

REFERENCES

Andrews, M. D. (1997, May/June). What matters most for teacher educators. *Journal of Teacher Education, 48*, 167-176.

Barufaldi, J. P. (1977). Realities of the times in science education. In G. Hall (Ed.), *AETS yearbook: Science teacher education vantage point 1976* (p. 18). ERIC: Ohio: Columbus, Ohio State University.

Barufaldi, J. P. (1987, April). *Perspectives in research in science education: A legacy and a promise.* Presidential address at the meeting of the National Association for Research in Science Teaching, Atlanta, GA.

Barufaldi, J. P. (1998, June). *The nature and process of collaboration to improve the teaching and learning of science.* Paper presented at the Science and Mathematics Conference, University of Lisbon, Portugal.

Barufaldi, J. P., & Reinhartz, J. (1998). *Collaborative driven professional development: Models in science education.* Unpublished concept paper. Austin: The University of Texas at Austin, College of Education.

Bybee, R. W. (1993). *Reforming science education: Social perspectives and personal reflections.* New York: Teachers College Press.

Cobb, P., & Yackel, E. (1996). Constructivist, emergent, and sociocultural perspectives in the context of development research. *Educational Psychologist, 31*, 175-190.

Darling-Hammond, L. (1994, February). *Standards for teachers.* Thirty-fourth Charles W. Hunt Memorial Lecture presented at the American Association for Colleges of Teacher Education meeting, Chicago, IL.

Darling-Hammond, L. (1996). The quiet revolution: Rethinking teacher development. *Educational Leadership, 53,* 4-10.

Division of Educational System Reform/National Science Foundation. (1997). *Instrument for annual report of progress in systemic reform.* Washington, DC: Author.

Driver, R., & Easley, J. (1978). Pupils and paradigms: A review of literature related to concept development in adolescent science students. *Studies in Science Education, 5,* 61-84.

Gardner, D. P. (1983). *A nation at risk: The imperative for educational reform.* Washington, DC: U.S. Government Printing Office.

Gilbert, J. (1994). The construction and reconstruction of the concept of the reflective practitioner in the discourses of teacher professional development. *International Journal of Science Education, 16,* 511-522.

Goertz, M. E., Floden, R. E., & O'Day, J. (1996). *Studies of education reform. Systemic reform. Volume I: Findings and conclusions.* Washington, DC: U.S. Government Printing Office.

Hargreaves, A. (1995). Development and desire: A postmodern perspective. In T. R. Guskey & M. Huberman (Eds.), *Professional development in education: New paradigms and practices* (pp. 9-34). New York: Teachers College.

Hatano, G. (1993). Time to merge Vygotskian and constructivist conceptions of knowledge acquisition. In E. Forman, N. Minick, & C. Stone (Eds.), *Contexts for learning* (pp. 153-166). New York: Oxford University Press.

Howe, A. C., & Stubbs, H. S. (1996). Empowering science teachers: A model for professional development. *Journal of Science Teacher Education, 8,* 167-182.

Hurd, P. H. (1997). *Inventing science education for the new millennium.* New York: Teachers College Press.

Jbeily, K. (1998). *Achievements during 1996-97 of the Texas regional collaboratives for excellence in science teaching.* Unpublished paper. Austin: The University of Texas at Austin, College of Education.

Jbeily, K., & Barufaldi, J. P. (1998). *Texas regional collaboratives for excellence in science teaching.* Unpublished paper. Austin: The University of Texas at Austin, College of Education.

Lieberman, A. (1996). Creating intentional learning communities. *Educational Leadership, 54,* 51-55.

Loucks-Horsley, S., Stiles, K., & Hewson, P. (1996 May). *Principles of effective professional development for mathematics and science education: A synthesis of standards.* University of Wisconsin-Madison: National Center for Improving Science Education.

Mattessich, P. W., & Monsey, B. R. (1992). *Collaboration: What makes it work—A review of literature on factors influencing successful collaboration.* St. Paul, Minnesota: Amherst H. Wilder Foundation.

Miles, M. (1995). New paradigms and practices, Foreword. In T. Guskey & M. Huberman (Eds.). *Professional development in education: New paradigms and practices* (pp. vii-ix). New York: Teachers College Press.

National Research Council (NRC). (1996). *National science education standards.* Washington, DC: National Academy.

National Science Board. (1983). *Educating Americans for the 21st century.* Commission on Precollege Education in Mathematics, Science, and Technology. Washington, DC.: Author.

Smylie, M., & Conyers, G. (1991). Changing conceptions of teaching influence the future of staff development. *Journal of Staff Development, 12,* 12-16.

Wasley, P. A. (1991). *Teachers who lead: The rhetoric of reform and the realities of practice.* New York: Teachers College Press.

SECTION TWO

SPECIAL ISSUES-DRIVEN MODELS

INSTRUCTIONAL CONGRUENCE TO PROMOTE SCIENCE LEARNING AND LITERACY DEVELOMENT FOR LINGUISTICALLY DIVERSE STUDENTS

Okhee Lee[1] & Sandra H. Fradd[2]
[1] *University of Miami*, [2] *University of Florida*

INTRODUCTION

The vision of current standards-based reform promotes high academic standards for all students. In science education, the key principle guiding the *National Science Education Standards* [NSES] (National Research Council [NRC], 1996) is: "*Science is for all students.* This principle is one of equity and excellence" (p. 20, original emphasis). With increasing numbers of students from diverse backgrounds, the task involves simultaneously considering students' language and cultural experiences and the demands of challenging academic standards (Darling-Hammond, 1996; McLaughlin, Shepard, & O'Day, 1995; Thompson, Wilder, & Atwater, this volume).

Although major reform documents in science education lay out what constitutes challenging science standards (American Association for the Advancement of Science [AAAS], 1989, 1993; NRC, 1996), they do not provide an implementation plan with specific strategies or procedures to achieve equity (Collins, 1998). The knowledge base to promote science learning for all students, including those from diverse languages and cultures, is limited (Atwater, 1994; Lee & Fradd, 1998). This knowledge base is essential, considering that nationally the school-aged population of English-only students decreased by nearly 8% during the past decade, while the population of students learning English as a new language increased by almost 40% (U.S. Bureau of the Census, 1993; U.S. Bureau of the Census, 1999).

We propose a model, "instructional congruence", to enable linguistically diverse students to achieve high academic standards in science and literacy. The purpose of the chapter is three-fold: (a) to explain the framework of instructional congruence, (b) to illustrate its key aspects using examples of classroom practices, and (c) to offer practical suggestions for establishing instructional congruence. First, we explain the theoretical framework for the instructional congruence model. Second, we describe professional development activities enabling elementary teachers who shared the languages and cultures of their students to establish instructional congruence in classrooms that served as research sites. Third, we describe the research and classroom contexts in which the model was conceptualized and implemented. Fourth, we provide examples of how the teachers established instructional congruence to promote students' science learning and literacy development. Finally, we briefly present student achievement results indicating the effectiveness of the model and offer im-

D.R. Lavoie and W.-M. Roth (eds.), Models of Science Teacher Preparation, 109–126.

plications for preservice teacher education in making science available for all students.

INSTRUCTIONAL CONGRUENCE: THEORETICAL FRAMEWORK

With the increasing diversity of the student population, one of the challenges teachers face is how to enable students from diverse languages and cultures to acquire academic knowledge across subject areas (August & Hakuta, 1997; García, 1993). Although literacy development can be challenging for all students, the challenge can be particularly great for students developing literacy and English language proficiency as they gain academic knowledge. A comprehensive set of teaching practices is required to promote achievement in content areas, such as science, while simultaneously developing literacy and English language proficiency. In this section, we discuss a view of science and science learning that considers students' languages and cultures. We explain the instructional congruence model as a means for integrating science learning and literacy development for linguistically diverse students.

View of Science and Science Learning Considering Language and Culture

All students bring into the classroom their ways of looking at the world that are formed by their environments and personal experiences (Driver et al., 1994). In science classrooms, students are expected to learn a body of science knowledge, the inquiry process of generating science knowledge, rules of science discourse, and habits of mind in terms of the values, attitudes, and scientific worldviews. Traditionally, the ways of knowing in science are defined in the *Western* science tradition (AAAS, 1989, 1993; NRC, 1996). The emerging literature on multicultural science education indicates that students from diverse languages and cultures display alternative ways of knowing that are sometimes incompatible with the nature of science or the way science is taught in school (Atwater, 1994; Brickhouse, 1998; Calabrese Barton, 1998; Lee & Fradd, 1998; Rosebery, Warren, & Conant, 1992). This literature challenges the way of knowing in science as traditionally defined and advocates for more inclusive views of science and science learning that consider diverse languages and cultures.

As issues of diversity and equity have become critically important in science education, science and science education communities have engaged in heated debate on some of fundamental issues. These include questions about what counts as science, what should be taught in school science, and how to promote science learning for diverse students (Brickhouse, 1998; Calabrese Barton, 1998; Eisenhart, Finkel, & Marion, 1996; Good, 1993; Hodson, 1993; Loving, 1997; Matthews, 1994, 1998; Ogawa, 1995; Rodríguez, 1997; Siegel, 1997; Stanley & Brickhouse, 1994; Taylor, 1998; Williams, 1994).

Science has generally been taught to be consistent with the nature and practice of science as traditionally defined, with little consideration of alternative ways of knowing, thinking, and interacting in diverse languages and cultures (Hodson, 1993; Ogawa, 1995; Stanley & Brickhouse, 1994; Taylor, 1998). Studies on language and culture, on the other hand, have focused on mismatches between culturally and linguistically diverse students' backgrounds and the expectations of school, with little consideration of the nature of science (Atwater, 1994; Cobern & Aikenhead, 1998; Lee & Fradd, 1998; McKinley, Waiti, & Bell, 1992). With the increasing diversity of the student population, there has been an effort to consider issues of language and culture in the science teaching and learning process. This effort is particularly important in science, considering low science performance and under-representation of students from diverse backgrounds in science and related fields.

The Model of Instructional Congruence

We conceptualize the model as "instructional congruence", where teachers mediate the nature of academic content and inquiry with language and cultural experiences of diverse students (Lee & Fradd, 1998). For example, in teaching science and literacy for linguistically diverse students, the model of instructional congruence includes four components: students, teachers, science, and literacy. To provide effective science and literacy instruction for these students, teachers need to integrate knowledge of (a) the students' language and cultural experiences, (b) science learning, and (c) literacy development. Through instructional congruence, teachers make academic content and inquiry (e.g., science) accessible, meaningful, and relevant for diverse students (e.g., linguistically diverse students).

The framework considers science and literacy as broadly defined in standards documents. According to NSES (NRC, 1996) and Project 2061 (AAAS, 1989, 1993), the major components of science learning include (a) key science concepts as well as big ideas in terms of patterns of change, systems, models, and relationships ("knowing"); (b) science inquiry emphasizing students' asking questions and finding answers ("doing"); (c) science discourse and multiple representations using various written and oral communication formats ("talking"); and (d) scientific habits of mind in terms of the values, attitudes, and world views in science (for detailed descriptions, see Lee [1998] and Lee and Fradd [1998]).

Literacy development is based on the standards for students learning English (International Reading Association and the National Council of Teachers of English [NCTE], 1994), English as a new (Teachers of English to Speakers of Other Languages [TESOL], 1997) and foreign language (American Council on the Teaching of Foreign Languages [ACTFL], 1996). Literacy development includes social and academic discourse in formal and informal settings (Cummins, 1984; Fradd & Larrinaga McGee, 1994). Social language is characterized as interpersonal and dependent on the culture of the communication, such as tone of voice, facial expressions, body movements, turn taking, and other aspects of interactional styles. Academic language is characterized as linguistically complex and cognitively demanding. Academic language is also characterized as the language of school instruction where understanding depends on knowledge of academic content and genre.

Instructional congruence has four important features. First, instructional congruence emphasizes the role of teachers in the instructional process. The model of *instructional* congruence underscores what teachers do through instruction and assessment to meet students' learning needs. Particularly, we highlight the insights and practices of teachers who shared the languages and cultures of their students and also understood science. The research on cultural congruence indicates that when teachers and students share the same language and culture, they tend to interact in ways that promote the students' participation and engagement (Au & Kawakimi, 1994; Trueba & Wright, 1992). Although critically important, cultural congruence may sometimes be incompatible with the nature of science (Lee & Fradd, 1996a). To establish instructional congruence, teachers need to integrate knowledge of students' languages and cultures with the nature of science, particularly when the two are incompatible.

Second, by specifically considering the nature of academic content and inquiry, such as science, this model provides "subject-specific" pedagogies for diverse students. Pedagogies based on specific cultural patterns are alternately termed "culturally relevant", "culturally appropriate", "culturally congruent", "culturally responsive", and "culturally compatible" (for a summary, see Ladson-Billings, 1995). While these pedagogies tend to focus more on language and cultural aspects of teaching, instructional congruence gives equal emphasis to the nature of academic content and inquiry and issues of language and culture. We use the model to indicate *congruence* with both academic content and students' language and culture.

Third, instructional congruence promotes student learning in both *science and literacy*. Although science and literacy in the school curriculum appear distinct subject areas, they are closely related in terms of students' development of academic discourse required to meet academic standards. Instruction in each area holds the potential to enhance students' achievement in the other (Casteel & Isom, 1994; Lee & Fradd, 1996b). For example, hands-on science activities in collective settings encourage students to engage in both social and academic discourse. Hands-on science activities also promote students' movement from concrete to abstract thinking. Language functions, such as describing, predicting, reflecting, and integrating (Fradd & Larrinaga McGee, 1994; Tough, 1986) are also important science inquiry skills (Casteel & Isom, 1994). Literacy development involves communication in a variety of formats, such as orally, in writing and drawing, and through the use of tables and graphs. Multiple representational formats are also ways of presenting and reporting ideas in science. Thus, effective science instruction for linguistically diverse students also promotes literacy development and English language learning in a systematic and explicit manner.

Finally, *constructivism* is at the core of instructional congruence. With students from diverse languages and cultures, the notion of personal constructivism needs to be extended to sense making in the contexts of students' languages and cultures (Cobern, 1993; Cobern & Aikenhead, 1998). While science education has often ignored the language and cultural experiences that students from diverse backgrounds bring to the classroom (Gallard, 1993), instructional congruence capitalizes on these experiences as a valuable resource for teaching and learning. Instructional congruence enables students to construct knowledge of academic content and inquiry within their language and cultural contexts (Aikenhead, 1996; Cobern & Aikenhead, 1998; Gallard, 1993; Taylor, 1998). Thus, instructional congruence enhances students' competence and appreciation of their languages and cultures, in addition to promoting academic achievement in science and literacy (Cochran-Smith, 1995; Darling-Hammond, 1996; Ladson-Billings, 1995; Moll, 1992).

We have developed the model of instructional congruence through a series of research projects. Although specific research questions and objectives differ among these projects, the overall goal has been to determine the most effective ways in promoting science learning and language development for linguistically diverse students. In this chapter, we focus specifically on one project (Fradd, Lee, & Sutman, 1995-1998). We describe the model of instructional congruence as a way to promote science learning and literacy development simultaneously for linguistically diverse students. We highlight the role of teachers who share the language and culture of their students in establishing a knowledge base for instructional congruence.

TEACHER PROFESSIONAL DEVELOPMENT IN ESTABLISHING INSTRUCTIONAL CONGRUENCE

Approximately 20 teachers participated over the three-year period of research. This chapter focuses on six Hispanic and two Haitian teachers who shared the same language and cultural backgrounds of their students. The Hispanic group included one male and five female teachers. The Haitian group included one male teacher and one female teacher. All were born outside the U.S. and were fluent in English and either Spanish or Haitian Creole. Their teaching experiences ranged from three to over 12 years. Two of the Hispanic and one of the Haitian teachers made mid-career changes into teaching. Their principals recommended them for their teaching excellence and commitment to their students. Their participation in the research was from one to three years. Participation was voluntary and no compensation was made in terms of money or graduate credits.

Teacher professional development was a major aspect of the research. During the spring semester of the first year, we focused on observations of science instruction and teacher-student interactions in natural classroom settings, with limited support for professional development. During the second and third years, the teachers engaged in professional development activities and had extensive interactions with the project personnel. Because an important purpose of the research was to gain insights from capable and committed teachers who related science to their students' language and cultural experiences, we did not impose specific instructional approaches on the teachers.

Professional development activities occurred in multiple ways. We held four full-day workshop meetings each year. We also engaged in extensive conversations with teachers in small groups and individually. We met with teachers at each school site to address their needs and concerns related to practices and policies for science and literacy instruction. After classroom observations, we spent brief moments with teachers to gather their feedback and insights about the lessons. At these formal and informal meetings, various issues were discussed ranging from instructional strategies and insights, science education and TESOL standards documents, school and district policies for instruction and assessment, revision of the instructional units, and student achievement outcomes. One of the most frequently discussed themes was the application of theoretical frameworks, including cultural and instructional congruence, to classroom teaching.

The areas of emphasis for professional development evolved over the years. Decisions were made mutually by the project personnel and the teachers. Although initially the project personnel organized the meetings, teachers gradually increased their level of involvement. Over time, collaboration between the teachers and the project personnel was formed to share insights, reflections, and suggestions (Fradd et al., 1997). At the beginning of the research, all of the teachers indicated lack of confidence in science and science instruction and many expressed apprehension and dislike of science. In response to the teachers' request, the initial area of emphasis involved science. During the workshop meetings, the teachers engaged in the science activities they would be using with students, discussed key science concepts and big ideas, and shared teaching strategies.

As the teachers became more familiar with science, they began to consider ways to relate science to their students' language and cultural experiences. Although the teachers were aware of students' language and cultural experiences, they had not actively considered how to incorporate language and culture to enhance instruction. Discussion of language and culture involved the teachers' reflections on personal experiences as new arrivals in the United States, learning English as a new language, and affirmation of their cultural identities. As they recognized the importance of shared languages and cultures, they explicitly integrated this knowledge in instruction. Because of the immediate concern with science and, then, language and culture, literacy did not become a focus for quite some time. The teachers considered how to promote literacy as part of science instruction. Gradually, they discussed how cultural congruence could both promote and hinder science learning and how they could move beyond cultural congruence into instructional congruence. The teachers realized that the nature of science was sometimes incompatible with students' cultural experiences and expectations for classroom participation. This insight enabled the teachers to recognize the challenges of integrating science, students' language and culture, and literacy in ways that were meaningful and relevant for their students.

THE RESEARCH CONTEXT

Student Participants

The research was conducted in a large metropolitan school district in the Southeast, In this district, a high proportion of students are from non-English language backgrounds. During the 1997-98 school year (the third and final year of the research), the ethnic composition of the district was 51% Hispanic, 34% Black (including Haitian students), 13% White Non-Hispanic, and 2% Asian and Native American students. More than 30% of elementary students were enrolled in English to Speakers of Other Languages (ESOL) programs.

The research involved fourth-grade students at six school sites, each representing the dominant language of the community served by the school. These schools included two predominantly bilingual Hispanic, two bilingual Haitian, and two monolingual English-speaking groups. Fourth-grade was selected for important reasons. By the fourth grade, literacy becomes an important means to learn academic content and inquiry, beyond simply learning to read and write (Ruddell & Ruddell, 1994). Compared to earlier grade levels, science learning also becomes more demanding as students are expected to learn science content and inquiry.

Science and Literacy Instruction

We developed two instructional units ("The Water Cycle" and "Weather") to provide effective instruction for students learning English as a new language. The water cycle (and changes of state) and weather were two key topics in the district science curriculum at fourth grade. Although materials development was not part of the project, it became necessary because we could not locate effective instructional materials to promote both science and literacy for students learning English as a new language. The science and literacy components in the two units were based on standards documents in science (AAAS, 1989, 1993; NRC, 1996) and literacy (ACTFL, 1996; IRA & NCTE, 1994; TESOL, 1997).

While changes of state are more readily controlled within the classroom setting, weather phenomena involve multiple variables interacting to produce a variety of weather conditions. The water cycle unit focused on changes of state including melting, freezing, evaporation, and condensation. The unit concluded with a lesson on the water cycle that integrated all the concepts presented throughout the unit. Building on the water cycle unit, the weather unit presented the concepts of temperature, wind, humidity, clouds, rain, air pressure, air masses, and fronts. The unit concluded with a lesson on hurricanes that integrated major weather concepts. Using weather instruments, students measured weather conditions in local settings. Students also learned to read and explain weather maps in local newspapers. The two units focused on "big ideas", including patterns of change, systems, models, and relationships. Students learned these science concepts and big ideas through hands-on science inquiry. Both units emphasized the importance of relating science to students' language and cultural experiences. Both units also emphasized promoting students' literacy and English language proficiency.

As a part of the research, teachers were provided with science supplies and equipment, in addition to student copies of the units. To relate science to students' everyday experiences, most of the supplies were inexpensive, household items. Because of easy access to household items, teachers sometimes told the students about activities with their families at home. Students also volunteered to do activities they could do at home on their own. For example, as an extension of a lesson on comparing the weight and volume of water before and after freezing, students did a freezing activity with other liquids at home. In class, they discussed variations in results with different liquids. Throughout the weather unit students learned to read weather maps on local newspapers in class. As an extension of the class activity, students recorded weather information from television at home and discussed the information in class. Such home-extension activities encouraged participation of parents and families in children's school learning. Although important for all students, these types of extensions were important in enabling families with little experience in science to become familiar with the science content taught at school.

During the first year, instruction of one unit continued from February through April. During the second and third years, instruction of two units continued from September through February. Within the overall framework of instructional congruence, teachers were encouraged to provide instruction in ways that they considered effective with their students based on shared understandings of language and culture.

Data Collection and Analysis

Multiple sources of data were collected with teachers. First, we visited the classrooms, on average, twice a week with every teacher. We took fieldnotes and videotaped selected lessons with each teacher. Second, we conducted formal interviews with individual teachers once each year. The interviews were transcribed verbatim. Third, we took notes about teachers' participation, feedback, insights, and reflections at formal meetings. We videotaped some of the meetings. Finally, we recorded formal and informal conversations with small group and individual teachers.

With the eight teachers who were the focus of this chapter, we analyzed the data using qualitative methods (Erickson, 1986; Miles & Huberman, 1994; Strauss & Corbin, 1990). For each component of science learning emphasized in the research, we identified major patterns and themes with regard to teachers' understandings and practices in relating science with students' language and culture. We also identified vignettes to illustrate the patterns and themes. Multiple data sources allowed triangulation of interpretations.

INSTRUCTION TO ESTABLISH INSTRUCTIONAL CONGRUENCE

Each component of science learning is described in terms of its importance for linguistically diverse students and is accompanied by examples of how teachers established instructional congruence. The examples highlight: (a) how teachers provided science instruction based on their students' language and cultural experiences and (b) how teachers promoted English language proficiency along with science.

Knowing Science

Knowing science involves making meaning of scientific knowledge and vocabulary. Although prior knowledge and personal experience are important in acquiring new knowledge for all learners (Driver et al., 1994; Posner et al., 1982), building on already established understandings is particularly relevant for linguistically diverse students with limited formal experience with science. The use of culturally familiar examples, analogies, and contexts promotes science learning with students from diverse backgrounds (Barba, 1993; Cobern & Aikenhead, 1998; Ladson-Billings,

1994, 1995; Lee & Fradd, 1998; Matthews & Smith, 1994). These students often come from environments in which science does not play a key role. In addition, their ways of knowing may be incompatible with the nature of science as it has traditionally been defined and taught in school. Yet, the students bring personal knowledge and experience that can be valuable in learning science. The following examples illustrate the importance of language and cultural experiences in science learning.

During a lesson on evaporation, a Hispanic teacher asked the class to explain humidity. Most of the students were in the process of learning English as a new language. In response to the teacher's question, a few students raised hands to volunteer their answers. Then, the teacher rephrased the question, asking, "When your father went outside this morning and he said *humedo* [humid in Spanish]. What does that mean?" Immediately, a number of students waved their hands, and some even jumped out of their seats with excitement. The teacher waited a few seconds to calm down the class, and called on students to give their responses. Students quickly made the connection between water vapor in the air and humidity. In this example, the word humidity in English might have been foreign to the students. In contrast, the words *humedo* and *humedad* (humidity) in Spanish within a culturally relevant context connected to their experiences and emotions, triggering enthusiastic reactions from the students. By relating to students' language and cultural experiences, the teacher promoted their understanding of science and English language learning.

As another example, a lesson in the weather unit presented the concept that the earth is heated unevenly because the sun's rays are more direct on some places than others. Using a world map, a teacher asked the class about differences in temperature among five major cities in North America. Although the lesson was planned to stop there, the teacher extended the discussion to include the Caribbean and South America, the birthplaces of many of the students. He selected five capital cities and asked the class about temperature differences. He asked the class to identify patterns in temperature differences among the 10 cities in the Americas. The teacher made sure the students understood the concept—the cities closer to the equator tend to be warmer because the sun's rays reach these places more directly. Then, several students asked questions as they related the lesson to their prior knowledge and personal experiences. One student asked, "Colombia is close to the equator, but when I visit my grandma in Bogotá in summer, it is cool there. Why?" Another student asked why a city close to the equator sometimes got cool and chilly. Through this discussion, the students recognized that other factors, such as elevation, proximity to the ocean, and frontal activities, were also important in determining temperatures. As they learned science concepts about weather and big ideas of patterns and systems, they also related science to their own cultural backgrounds. By utilizing the cultural experiences that students brought to the classroom, the teacher made science relevant and meaningful for the students.

Doing Science

As a way of knowing science, students engage in inquiry, or doing science, by manipulating materials, making observations, proposing explanations, interpreting and verifying evidence, and constructing ideas. NSES (NRC, 1996) emphasized that inquiry is at "the heart of science and science learning" (p. 15) and "inquiry into authentic questions generated from student experiences is the central strategy for teaching science" (p. 31).

Hands-on inquiry activities are particularly effective for students from diverse backgrounds and students learning English as a new language. Students' interest and curiosity in hands-on activities prompt them to communicate their observations and ideas with others. The physical presence of hands-on activities also reduces the burden of language demands, enabling students to concentrate on meanings of the activities. Such activities provide opportunities for students to move from concrete experiences to abstract thinking. In addition, hands-on activities in group settings are generally consistent with the norms and values of collaboration, sharing, and support in diverse languages and cultures (Atwater, 1994; Lee, Fradd, & Sutman, 1995).

Despite strengths and advantages, science inquiry poses challenges to students from diverse backgrounds. These students often come from cultures that respect teachers' authority in telling and directing students, rather than promoting students' exploration or alternative solutions (Atwater, 1994; McKinley, Waiti, & Bell, 1992; Ogunniyi, 1988; Prophet & Rowell, 1993). Because inquiry and questioning is not part of their cultural experience, these students need to learn to engage in science inquiry as they learn to ask questions and find answers. They also need to learn to question and argue with the teacher and peers in the science classroom. Indeed, the need for explicit instruction in the context of meaningful and relevant tasks and activities for students from diverse backgrounds has been advocated in literacy instruction (Delpit, 1988; Reyes, 1992), classroom discourse (Gee, 1997, in press), and science instruction (Fradd & Lee, 1999).

The Hispanic and Haitian teachers in our research initially used explicit, teacher-directed interactions. They orchestrated the class as a whole. Even when students worked in small groups, the teachers involved the small groups as parts of the whole class. One teacher expressed her insights as follows:

> The students who need to do inquiry the most are those who also need to learn how to observe, measure, and compare things. These students have no idea what to look at, how to hold an instrument, or how to record data in a chart.... I think that students have to start with simple, concrete activities and basic experiences and teachers have to be in control in order for the students to be successful.

As instruction progressed, the teachers gradually made the transition from a teacher-directed mode of instruction to one where students exercised a degree of autonomy. Although the extent of teacher control and explicitness began to decrease, the teachers continued to ensure that everyone participated and understood as lessons proceeded smoothly and efficiently. Because teachers made sure students conducted activities consistently and obtained measurements accurately, students engaged in simple but meaningful discussions. During these times, many of the students were learning how to voice their ideas and interpretations of their observations. It took a great deal of students' practice with teachers' guidance until the students could ask questions and engage in more open-ended activities and discussions (Fradd & Lee, 1999).

Talking Science

Although knowing and doing science have long been emphasized in science learning, recent reform highlights the importance of discourse and communication (Gee, 1996, in press; Lemke, 1990; Palincsar, Anderson, & David, 1993). The NSES (NRC, 1996) states, "[Teachers] structure and facilitate ongoing formal and informal discussion based on a shared understanding of rules of scientific discourse. A fundamental aspect of a community of learners is communication" (p. 50).

Talking science is critically important for students learning English as a new language. Through oral and written communication with other students and teachers as well as printed and visual materials, students develop literacy and English language proficiency as well as learn science content. Although much of the communication in science has been done through reading and writing, oral discourse is being emphasized for students with limited literacy and language development (Lee & Fradd, 1996b; Yore, Holliday, & Alverman, 1994). In addition, these students can express their ideas and understandings through multiple representational formats, such as spoken, written, pictorial, graphic, mathematical, and electronic communication. As the students develop their literacy and science knowledge, they gradually learn to use representational formats that are effective for particular tasks.

For example, a science activity for condensation involved observing the differences on the outside of the two cups, one with cold water and the other with tap water. Students were asked to describe what they saw above and below the water line of each cup. Recognizing the students' difficulty following the directions, a teacher drew on the board the two cups with positional words for each cup, "inside/outside" and "below/above". In this example, although words like "inside/outside" or "below/above" are generally used as social language in everyday conversation, these positional words became academic language because of their importance in successfully completing this academic task. For students learning English as a new language, the teacher helped the students develop literacy and English language proficiency. In addition, the teacher helped the students acquire the skills in following science procedures and the dispositions of being accurate and systematic in observations.

Scientific Habits of Mind

Science involves certain values, attitudes, and worldviews (AAAS, 1989, 1993; NRC, 1996). Some scientific values and attitudes are found in most cultures, such as curiosity, interest, insight, wonder, diligence, and persistence. Others are more characteristic of Western science, such as being independent, critical, skeptical, and argumentative as well as questioning rather than deferring to authority. In addition, scientific values and attitudes include teamwork, collaboration, and shared responsibility for learning with others. Science also involves a scientific worldview that seeks to understand how the world works through observation, experimentation, and validation. The scientific worldview differs from alternative views, such as personal beliefs, myths, spirits, superstition, and supernatural forces.

Some of the scientific values and attitudes that promotes "a critical and questioning stance" (Williams, 1994) may be incongruent with the norms of diverse cultures that favor cooperation, social support, consensus building, and respect for authority (Atwater, 1994; Hodson, 1993; Lee & Fradd, 1998). On the other hand, some aspects of these cultural values, particularly the notion of group collaboration, are consistent with team work and shared responsibility that are also important in a science learning community. Because values, attitudes, and worldviews are fundamental aspects of students' home and community lives, cultivation of scientific habits of mind requires a great deal of sensitivity and consideration.

Although teachers encouraged students to work individually and independently, they also valued teamwork and collaboration, as the students generally preferred it. While interacting in small groups, students would often work cooperatively and help each other. For example, as a culminating activity of the weather unit, several teachers asked the students to make class presentations on weather reports including the summary of the weather for the week and prediction for the next several days. The students generally preferred to do the activity in groups of two or three students, rather than individually. In one class, when a group of two students were hesitant in giving their report to the class, a girl encouraged the students by sharing her experience with the class that she was nervous before her group's presentation but everything went well. When the two students gave their report, the class applauded and the two students smiled to the class in return.

DISCUSSION AND IMPLICATIONS

Teachers play a critical role in establishing instructional congruence. Through instructional congruence, students achieved in both science and literacy. Evaluation of the instructional congruence model is discussed in terms of student outcomes. Then, implications for preservice teacher education are discussed.

Evaluation of the Model: Student Achievement

Several assessments were used with students in science and literacy (see Fradd & Lee, 2000). First, during the second and third years, we administered project-developed tests on the water cycle and weather units with all students prior to and after instruction. Second, for more in-depth examination, elicitations were conducted with randomly selected student dyads to assess science knowledge, literacy, cognitive strategy use, and reflections on learning. Finally, in addition to written samples of elicitation students, written samples of science activities were obtained in selected classrooms.

As an overall measure of the impact of instructional congruence on students, only the science achievement results on the two unit tests with all students are briefly reported here (for more detailed descriptions, see Fradd & Lee, in press). The aggregated total score for the water cycle and weather unit was 132 points. During the second year, the mean score on the combined water cycle and weather pre-tests was 20.16 (*SD* = 8.89) and the mean score on the combined post-tests was 47.65 (*SD*

= 12.38). The students performed significantly better on the post-tests than pre-tests (t = 27.33, $p < .001$). During the third year, the mean score on the pre-tests was 24.48 (SD = 11.13) and the mean score on the post-tests was 60.00 (SD = 15.50). Again, the students performed significantly better on the post-tests than pre-tests (t = 28.83, $p < .001$).

The results indicate that when the teachers mediated science with students' language and cultural experiences, the students made sense of science in ways that were culturally and linguistically meaningful and relevant as well as scientifically accurate (Cobern & Aikenhead, 1998; Gallard, 1993). While learning science, the students in the research also increased their levels of literacy and English language proficiency. In addition to their achievement in science and literacy, the students recognized the value of their own language and cultural experiences (Ladson-Billings, 1995; Moll, 1992). Thus, instructional congruence enabled these students to become more knowledgeable about their languages and cultures (Lee & Fradd, 1998; McKinley, Waiti, & Bell, 1992).

Implications for Preservice Teacher Education

One of the difficulties of teacher education involves the issue of culture in subject area instruction. Particularly, mathematics and science are often misunderstood as "culture free" (Banks, 1993; Lee, 1999; Peterson & Barnes, 1996; Secada, 1989). Because mathematics and science tend to be presented as a set of objective and universal facts and rules, they are not considered as socially and culturally constructed disciplines. Thus, understanding the nature of academic disciplines, such as mathematics and science, requires that teachers change their beliefs about what counts as knowledge and how knowledge is related to diverse languages and cultures (Lee, 1999; Secada, 1989). Understanding of diverse languages and cultures also requires that teachers assess their own identities and recognize how students' identities may interact with science learning (Banks, 1993). Such analysis can lead teachers to make fundamental transformations in their beliefs and practices (Cochran-Smith, 1995; Ladson-Billings, 1994; Valli, 1995). Although some teachers benefit from self-analysis and reflection as they become more aware and understanding of diversity, others become less tolerant of diversity. Because of its potentially contentious nature, some educators may simply consider the issue of language and culture "too hot to handle" (Peterson & Barnes, 1996, p. 489).

The process of establishing instructional congruence with teachers who shared the language and culture as their students can provide important insights for teachers who do not share students' ethnolinguistic backgrounds. The establishment of instructional congruence occurs gradually as teachers construct their understandings of science and relate these with students' learning needs. By integrating instruction in science and literacy, teachers can make science relevant and meaningful, as they also promote language learning and proficiency in English with ethnolinguistically diverse students. Based on our research and relevant literature, we offer the following suggestions for preservice teacher education to establish instructional congruence in science classrooms.

Preservice science teachers need to become familiar with the expectations for science teaching and learning presented in standards documents (AAAS, 1989, 1993; NRC, 1996). To relate science to students' languages and cultures, teachers encourage students to share their prior knowledge and experiences related to science topics. In communicating with diverse students, teachers consider multiple ways of representing ideas within the context of diverse languages and cultures. Teachers accept students' ways of communication and interaction as appropriate, while modeling new ways of participating in science. Teachers also accept students' cultural values, attitudes, and worldviews as valid, while demonstrating the worldview of the science community.

In relating science to students' language and cultural experiences, teachers need to be aware that although cultural congruence can facilitate science learning, it may sometimes be incompatible with science as presented in standards documents. Thus, teachers identify specific instructional practices that are congruent with students' language and cultural experiences, as well as specific practices that may limit students' participation in science. Initially, teachers provide extensive instructional scaffolding for students to develop basic concepts and skills in science. As students gain knowledge and experience, teachers gradually make a transition into instructional congruence.

With ethnolinguistically diverse students, literacy and language development is an integral part of any subject area instruction, including science. Preservice elementary teachers need to be familiar with the expectations for literacy and language proficiency in English and other languages presented in standards documents (ACTFL, 1996; IRA & NCTE, 1994; TESOL, 1997). Teachers integrate literacy and language development as part of science instruction by providing opportunities for students to communicate their ideas in oral and written modes. To provide effective linguistic scaffolding, teachers adjust language requirements in learning science by encouraging students to communicate their ideas in multiple representational formats, including charts, graphs, diagrams, and computer technology. Teachers also encourage students to communicate about science in their home languages as well as English. In making a transition into instructional congruence, teachers start with large and small group settings and gradually enable students to perform individually and independently.

SUMMARY

With increasing numbers of students from diverse language and cultural backgrounds, the nation is challenged to achieve the goal of high academic standards for all students. In this chapter we have presented a model of *instructional congruence* to promote linguistically diverse students' achievement in science. We have described how fourth-grade elementary teachers who shared the language and culture of their students established instructional congruence to promote students' science learning and literacy development. Instructional congruence occurred as teachers mediated the nature of academic content and inquiry with students' language and cultural experiences. A discussion of the implications instructional congruence for teacher enhancement concludes this chapter. With adaptations for specific learner needs, the model can be extended to other groups at the margins of science, including students from low socio-economic levels, students with disabilities, and female students. The practical knowledge of teaching by the individual teachers in the research can be incorporated as an important part of theoretical knowledge required for effective instruction (Ladson-Billings, 1994, 1995). This knowledge base can be shared with teachers from a variety of backgrounds to make science available for all.

NOTES

The authors recognize the support from the National Science Foundation under Grant No. REC-9552556. Any opinions, findings, and conclusions or recommendations expressed in this publication are those of the authors and do not necessarily reflect the position, policy, or endorsement of the funding agency.

REFERENCES

Aikenhead, G. S. (1996). Science education: Border crossing into the subculture of science. *Studies in Science Education, 26,* 1-52.

American Association for the Advancement of Science (1989). *Science for all Americans.* New York: Oxford University Press.

American Association for the Advancement of Science (1993). *Benchmarks for science literacy.* New York: Oxford University Press.

American Council on the Teaching of Foreign Languages [ACTFL]. (1996). *Standards for foreign language learning: Preparing for the 21st century.* Lawrence, KS: Allen Press.

Atwater, M. M. (1994). Research on cultural diversity in the classroom. In D. L. Gabel (Ed.), *Handbook of research on science teaching and learning* (pp. 558-576). New York: Macmillan.

Au, K. H., & Kawakimi, A. J. (1994). Cultural congruence in instruction. In E. R. Hollins, J. E. King, & W. C. Hayman (Eds.), *Teaching diverse populations: Formulating a knowledge base* (pp. 5-24). Albany, NY: State University of New York Press.

August, D., & Hakuta, K. (Eds.). (1997). *Improving schooling for language-minority children: A research agenda.* Washington, DC: National Academy Press.

Banks, J. (1993). Multicultural education: Historical development, dimensions and practice. In L. Darling-Hammond (Ed.), *Review of Research in Education, Vol. 19* (pp. 3-49). Washington, DC: American Educational Research Association.

Barba, R. H. (1993). A study of culturally syntonic variables in the bilingual/bicultural science classroom. *Journal of Research in Science Teaching, 30*(9), 1053-1071.

Brickhouse, N. (1998). Feminism(s) and science education. In B. Fraser & K. Tobin (Eds.), *International handbook of science education: Part II* (pp. 1067-1081). Dordrecht, the Netherlands: Kluwer Academic Publishers.

Calabrese Barton, A. (1998). Teaching science with homeless children: Pedagogy, representation, and identity. *Journal of Research in Science Teaching, 35,* 379-394.

Casteel, C. P., & Isom, B. A. (1994). Reciprocal processes in science and literacy learning. *The Reading Teacher, 47,* 538-545.

Cobern, W. W. (1993). Contextual constructivism: The impact of culture on the learning and teaching of science. In K. Tobin (Ed.), *The practice of constructivism in science education* (pp. 51-69). Washington, DC: American Association for the Advancement of Science.

Cobern, W. W., & Aikenhead, G. S. (1998). Cultural aspects of learning science. In B. Fraser & K. Tobin (Eds.), *International handbook of science education: Part I* (pp. 39-52). Dordrecht, the Netherlands: Kluwer Academic Publishers.

Cochran-Smith, M. (1995). Color blindness and basket making are not the answers: Confronting the dilemmas of race, culture, and language diversity in teacher education. *American Educational Research Journal, 32,* 493-522.

Collins, A. (1998). National Science Education Standards: A political document. *Journal of Research in Science Teaching, 35,* 711-727.

Cummins, J. (1984). *Bilingualism and special education: Issues in assessment and pedagogy.* Boston: College-Hill Press.

Darling-Hammond, L. (1996). The right to learn and the advancement of teaching: Research, policy, and practice for democratic education. *Educational Researcher, 25*(6), 5-17.

Delpit, L. (1988). The silenced dialogue: Power and pedagogy in educating other people's children. *Harvard Educational Review, 58,* 280-298.

Driver, R., Asoko, H., Leach, J., Mortimer, E., & Scott, P. (1994). Constructing scientific knowledge in the classroom. *Educational Researcher, 23,* 5-12.

Eisenhart, M., Finkel, E., & Marion, S. F. (1996). Creating the conditions for scientific literacy: A re-examination. *American Educational Research Association, 33,* 261-295.

Erickson, E. (1986). Qualitative methods in research on teaching. In M. C. Wittrock (Ed.), *Handbook of research on teaching* (3rd ed., pp. 145-158). New York: Macmillan.

Fradd, S. H., & Larrinaga McGee, P. (1994). *Instructional assessment: An integrative approach to evaluating student performance.* Reading, MA: Addison-Wesley.

Fradd, S. H., & Lee, O. (1999). Teachers' roles in promoting science inquiry with students from diverse language backgrounds. *Educational Researcher, 28*(6), 21-27.

Fradd, S. H., & Lee, O. (2000). Needed: A framework for integrating standardized and informal assessment for students developing academic language proficiency in English. In J. V. Tenajero & S. Hurley (Eds.), *Literacy assessment of bilingual learners* (pp. 130-148). Boston, MA: Allyn & Bacon.

Fradd, S. H., Lee, O., Cabrera, P., del Rio, V., Leth, A., Morin, R., Ceballos, M., Santalla, M., Cross, L., & Mathieu, T. (1997). School-university partnership to promote science with students learning English. *TESOL Journal, 7,* 35-40.

Fradd, S. H., Lee, O., & Sutman, F. X. (1995-1998). *Promoting science literacy for all Americans, including culturally and linguistically diverse students: Keeping the promise* (National Science Foundation, Research on Teaching and Learning Program, #9552556). Coral Gables, FL: University of Miami.

Gallard, A. J. (1993). Learning science in multicultural environments. In K. Tobin (Ed.), *The practice of constructivism in science education* (pp. 171-180). Washington, DC: American Association for the Advancement of Science.

García, E. E. (1993). Language, culture, and education. In L. Darling-Hammond (Ed), *Review of Research in Education, Vol. 19* (pp. 51-98). Washington, DC: American Educational Research Association.

Gee, J. P. (1997). *Science talk: Language and knowledge in classroom discussion.* Paper presented at the annual meeting of the National Association for Research in Science Teaching, Oakbrook, IL.

Gee, J. P. (in press). Progressivism, critique, and socially situated minds. In C. Edelsky & C. Dudley-Marling (Eds.), *Progressive education: History and critique.* Urbana, IL: National Council for the Teaching of English.

Good, R. G. (1993). Editorial: The slippery slopes of postmodernism. *Journal of Research in Science Teaching, 30,* 427.

Hodson, D. (1993). In search of a rationale for multicultural science education. *Science Education, 77,* 685-711.

International Reading Association and the National Council of Teachers of English. (1994). *Standards for the assessment of reading and writing.* Newark, Delaware: International Reading Association (with National Council of Teachers of English).

Ladson-Billings, G. (1994). *The dreammakers: Successful teachers of African American children.* San Francisco, CA: Jossey-Bass Publishers.

Ladson-Billings, G. (1995). Toward a theory of culturally relevant pedagogy. *American Educational Research Journal, 32,* 465-491.

Lee, O. (1998). *Current conceptions of science achievement in major reform documents and implications for equity and assessment* (Research Monograph 12). Madison, WI: University of Wisconsin, National Institute for Science Education.

Lee, O. (1999). Equity implications based on the conceptions of science achievement in major reform documents. *Review of Educational Research, 69*(1), 83-115.

Lee, O., & Fradd, S. H. (1996a). Interactional patterns of linguistically diverse students and teachers: Insights for promoting science learning. *Linguistics and Education: An International Research Journal, 8,* 269-297.

Lee, O., & Fradd, S. H. (1996b). Literacy skills in science performance among culturally and linguistically diverse students. *Science Education, 80*, 651-671.

Lee, O., & Fradd, S. H. (1998). Science for all, including students from non-English language backgrounds. *Educational Researcher, 27*(3), 12-21.

Lee, O., Fradd, S. H., & Sutman, F. X. (1995). Science knowledge and cognitive strategy use among culturally and linguistically diverse students. *Journal of Research in Science Teaching, 32*, 797-816.

Lemke, J. (1990). *Talking science: Language, learning, and values.* Norwood, NJ: Ablex Publishing.

Loving, C. C. (1997). From the summit of truth to its slippery slopes: Science education's journey through positivist-postmodern territory. *American Educational Research Journal, 34*, 421-452.

Matthews, C. E., & Smith, W. S. (1994). Native American related materials in elementary science instruction. *Journal of Research in Science Teaching, 31*, 363-380.

Matthews, M. (1994). *Science teaching: The role of history and philosophy of science.* New York: Routledge.

Matthews, M. (1998). The nature of science and science teaching. In B. Fraser & K. Tobin (Eds.), *International Journal of Science Education, Part two* (pp. 981-1000). Dordrecht, the Netherlands: Kluwer Academic Publishers.

McKinley, E., Waiti, P. M., & Bell, B. (1992). Language, culture and science education. *International Journal of Science Education, 14*, 579-595.

McLaughlin, M. W., Shepard, L. A., & O'Day, J. A. (1995). *Improving education through standards-based reform: A report by the National Academy of Education Panel on Standards-based Education Reform.* Stanford, CA: Stanford University, National Academy of Education.

Miles, M. B., & Huberman, A. M. (1994). *Qualitative data analysis: A sourcebook of new methods* (2nd ed.). Newbury Park, CA: Sage Publications.

Moll, L. C. (1992). Bilingual classroom studies and community analysis: Some recent trends. *Educational Researcher, 21*(2), 20-24.

National Research Council. (1996). *National science education standards.* Washington, DC: National Academy Press.

Ogawa, M. (1995). Science education in a multiscience perspective. *Science Education, 79*, 583-593.

Ogunniyi, M. B. (1988). Adapting western science to traditional African culture. *International Journal of Science Education, 10*, 1-9.

Palincsar, A. S., Anderson, C. W., & David, Y. (1993). Pursuing scientific literacy in the middle grades through collaborative problem solving. *The Elementary School Journal, 93*, 643-658.

Peterson, P. L., & Barnes, C. (1996). Learning together: The challenge of mathematics, equity, and leadership. *Phi Delta Kappan, 77*, 485-491.

Posner, G. J., Strike, K. A., Hewson, P. W., & Gertzog, W. A. (1982). Accommodation of a scientific conception: Toward a theory of conceptual change. *Science Education, 66*, 211-227.

Prophet, R. B., & Rowell, P. M. (1993). Coping and control: Science teaching strategies in Botswana. *Qualitative Studies in Education, 6*, 197-209.

Reyes, M. (1992). Challenging venerable assumptions: Literacy instruction for linguistically diverse students. *Harvard Educational Review, 62*, 427-446.

Rodríguez, A. J. (1997). The dangerous discourse of invisibility: A critique of the National Research Council's National Science Education Standards. *Journal of Research in Science Teaching, 34*, 19-37.

Rosebery, A. S., Warren, B., & Conant, F. R. (1992). Appropriating scientific discourse: Findings from language minority classrooms. *The Journal of the Learning Sciences, 21*, 61-94.

Ruddell, R. B., & Ruddell, M. R. (1994). Language acquisition and literacy processes. In R. B. Ruddell, M. R. Ruddell, & H. Singer (Eds.), *Theoretical models and processes of reading (4th ed..*, pp. 83-103). Newark, DE: International Reading Association.

Secada, W. G. (1989). Educational equity and equality of education: An alternative conception. In W. G. Secada (Ed.), *Equity in education* (pp. 68-88). Philadelphia, PA: The Falmer Press.

Siegel, H. (1997). Science education: Multicultural and universal. *Interchange, 28*, 97-108.

Stanley, W., & Brickhouse, N. (1994). Multiculturalism, universalism, and science education. *Science Education, 28*, 387-398.

Strauss, A., & Corbin, J. (1990). *Basics of qualitative research: Grounded theory procedures and techniques.* Newbury Park, CA: Sage Publications.

Taylor, P. C. (1998). Constructivism: Value added. In B. Fraser & K. Tobin (Eds.), *International handbook of science education, Part two* (pp. 1111-1126). Dordrecht, Netherlands: Kluwer Academic Publishers.

Teachers of English to Speakers of Other Languages. (1997). *ESL standards for pre-K-12 students.* Alexandria, VA: Author.

Tough, J. (1986). *Talk two: Children using English as a second language in primary schools.* London: Onyx Press.

Trueba, H. T., & Wright, P. G. (1992). On ethnographic studies and multicultural education. In M. Saravia-Shore, S. Arvizu (Eds.), *Cross-cultural literacy: Ethnographies of communication in multiethnic classrooms* (pp. 299-338). New York: Garland.

U.S. Bureau of the Census. (1993). *1990 profiles of the foreign-born population, selected characteristics by place of birth* (CPH-L-148). Washington, DC: U.S. Department. of Commerce, Population Division.

U.S. Bureau of the Census. (1999). *United States populations estimates, by age, sex, race, and Hispanic origin, April1, 1990 to April 1, 1999* (PPL-57). Washington, DC: U.S. Department. of Commerce, Population Division. *http://www.census.gov/populations/estimates/nation/intfile1.txt*

Valli, L. (1995). The dilemma of race: Learning to be color blind and color conscious. *Journal of Teacher Education, 46*, 120-129.

Williams, H. (1994). A critique of Hodson's "In search of a rationale for multicultural science education". *Science Education, 78*, 515-519.

Yore, L. D., Holliday, W. G., & Alverman, D. E. (Eds.). (1994). The Reading-science learning-writing connection. [Special Issue]. *Journal of Research in Teaching Science, 31*, 873-1073.

GENDER EQUITY AND SCIENCE TEACHER PREPARATION

Léonie J. Rennie
Curtin University of Technology

INTRODUCTION

Science traditionally is a subject in which educational outcomes are inequitable with regard to gender and, hence, of concern to the education of science teachers. For the most part, and especially beyond elementary school, sex differences are most visible in patterns of participation rather than performance, and there are numerous reviews and analyses of data sets which examine a range of differences or offer interpretations of the reasons for them (e.g., AAUW, 1992; Baker, 1998; Baker & Scantlebury, 1995; Becker, 1989; Catsambis, 1995; Ditchfield & Scott, 1987; Fennema, 1987; Kahle, 1985, 1988; Kahle & Meece, 1994; Kahle, Parker, Rennie, & Riley, 1993; Lee & Burkam, 1996; Linn & Hyde, 1989; Mason, 1995; Murphy & Gipps, 1996; Parker, Rennie, & Fraser, 1996; Parker, Rennie, & Harding, 1995; Rennie & Parker, 1993; Shroyer, Smith, Borchers, & Wright, 1994; Sjøberg & Imsen, 1988). This aspect of gender and science education need not be reviewed again here, rather this chapter focuses on the meaning of gender equity in science classrooms and how it might be addressed in science teacher education. This chapter is based around a model (Kahle et al., 1993) identifying clusters of variables associated with gender differences in science-related attitudes, perceptions, classroom behavior and learning outcomes. Four perspectives for conceptualizing and interpreting gender equity in relation to this model are described. A socially critical perspective is identified as the most promising for dealing with teachers' and students' perspectives about gender equity in science teacher preparation in the context of the *National Science Education Standards* (NRC, 1996). Examples of implementing the model using this perspective and the possible consequences to consider are discussed.

WHAT DOES GENDER EQUITY MEAN IN SCIENCE EDUCATION?

In my work with teachers, my interest in gender equity has often become a topic. On several occasions someone has rather aggressively said something like, "I have boys and girls in my physics class and I want you to know that I treat them exactly the same way". The first time this happened my immediate response was, "But why would you want to do that? Surely the students in your class aren't all exactly the same?" Then I wondered why these teachers felt the need to confront me in this way. Perhaps they themselves felt "confronted", assuming that I was about to blame

D.R. Lavoie and W.-M. Roth (eds.), Models of Science Teacher Preparation, 127–147.

them in some way for perceived gender-based inequities and were anxious to point out that they were not responsible for them.

Whatever the reason for their action, these teachers and my response to them highlights a fundamental question: what is meant by gender equity in science education? Does it mean equal treatment in the classroom, equal opportunities to participate, or equal outcomes at the end? The adoption in most developed countries, during the 1990s, of "standards" for science education, that is, "those common outcomes which a society expects schools to achieve with its children" (Black, 1995, p. 1) puts the focus on outcomes which can be achieved by all students. The *National Science Education Standards* (NRC, 1996) emphasize opportunity and access in the context of science for all students, "regardless of age, gender, cultural or ethnic background, disabilities, aspirations, or interest and motivation in science" (p. 2). In the words of Collins (1995), "Equity permeates the *Standards*" (p. 32). This is clear throughout the document.

> The *Standards* assume the inclusion of all students in challenging science opportunities and define levels of understanding and abilities that all should develop. They emphatically reject any situation in science education where some people—for example, members of certain populations—are discouraged from pursuing science and excluded from opportunities to learn science. (NRC, 1996, p. 20)

For the purposes of this chapter, the important implication of a standards-based education is what it means for the treatment of students in the classroom. It means that if all students are to be given opportunities to achieve understanding of science, they must be treated according to their needs. Sometimes students' needs are related to group membership, such as their gender and their cultural group, or to their learning styles, but their needs as individuals are of critical importance. Thus, providing opportunities for equal access and equal outcomes, almost by definition, means different treatment.

There is a large body of research which documents difference between groups of students based on their biological sex, rather than their individual needs. For example, in their international review, Parker et al. (1995) referred to well-recorded differences between boys, on average, and girls, on average, in their interests in different science subjects, their levels of confidence in their abilities, the ways they interact with each other, and their preferred learning styles. These authors also noted sex-based differences in teachers' expectations of students' behavior in class, teacher-student interaction, and the advice students are given for choice of subjects and careers. These differences reflect culturally based gender-stereotypes about the expected behavior of boys and girls that exert a powerful influence on the ways in which they interact with others. These stereotypes make it easy to forget that there is much greater variability among boys and among girls on any one of these variables, than there is between boys as a group and girls as a group. Thus, treating boys and girls in different ways simply on the basis of stereotypes, rather than their individual needs, is inequitable. Instead, we need to promote classroom practice that allows all students, irrespective of their sex and their social and cultural backgrounds, access to the "same valued knowledge" (Reid, 1995, p. 13) in the science curriculum and supports them to achieve to their potential.

The inclusion of *all* students is a principle underlying the *Standards* and other curriculum statements for an outcomes-based education, but what does it mean in practice?

> Inclusivity means providing all groups of students, irrespective of educational setting, with access to a wide and empowering range of knowledge, skills and values. It means recognizing and accommodating the different starting points, learning rates and previous experiences of individuals or groups of students. It means valuing and including the understandings and knowledge of all groups. It means providing opportunities for students to evaluate how concepts and constructions such as culture, disability, race, class and gender are shaped. (Curriculum Council, 1998, p. 17)

The focus in this chapter is on inclusivity in terms of gender. Thus, in equitable science classrooms, students are provided with a wide range of learning activities and a variety of assessment tasks so they can learn and demonstrate their learning in ways which suit them as individuals. And they also are given opportunities to examine the myths and stereotypes about science and the people who do science, enabling them to challenge and re-vision science in ways which are more inclusive of themselves, not only with respect to their gender roles but in terms of their culture, race, relationship with their environment, and other personal characteristics and motivations.

A MODEL FOR CONSIDERING GENDER EQUITY IN SCIENCE TEACHER EDUCATION

If a gender-inclusive science education is to be achieved, how should gender equity be conceptualized and interpreted in science teacher education? The *National Science Education Standards* (NRC, 1996) advocate equity and excellence in science education with statements such as "considerations of equity are critical in the science teaching standards" (p. 4). The emphasis on, for example, "understanding and responding to individual student's interests, strengths, experiences, and needs" (p. 52) and "supporting a classroom community with cooperation, shared responsibility, and respect" (p. 52) suggest that gender equity must permeate the preparation of science teachers if the standards are to be implemented. It is not simply a matter of introducing a few neat little strategies to "take care" of gender equity. A more comprehensive approach is needed.

In this chapter, I use a socio-cultural model of the relationship between gender and science developed by Kahle et al. (1993; Rennie, Parker, & Kahle, 1996) from a synthesis of research in science education. This model, presented in Figure 1, names the clusters of variables involved in understanding the relationship between gender and science and traces out the links between them. Four of the clusters of variables are teachers' beliefs and attitudes, students' beliefs and attitudes, and, at the heart of the model, teachers' and students' behavior in the classroom. The model assumes these variables to be in dynamic relationship, and the nature of the interactions among them determine the observable student outcomes, both in the long and short terms. Thus the model allows gender-related differences in outcomes (like students' career choices) to be traced back to both teachers' and students' beliefs and attitudes about science and gender, and how teachers and students interact in the classroom.

In their turn, both teachers' and students' beliefs and attitudes are shown in the model to be embedded in the socio-cultural educational context. The model represents a continuous process of adjustment, so that yesterday's lesson becomes part of today's previous experience and today's beliefs, attitudes, and behaviors are interpreted in that context. Thus the model can be used to assist interpretation of events and issues to explain gender differences in science-related attitudes, perceptions, classroom behavior and learning outcomes.

Although the Kahle et al. (1993) model maps out in a concrete way the clusters of variables which determine whether science education is gender equitable, it does not imply a particular perspective for explaining the relationships among them. In fact the model can be interpreted from several perspectives. This is a very important point, because the choice of strategies to address gender differences depends on how those differences are explained.

Like any other issue in science education, ways of thinking about gender and gender differences have changed over the years. We can trace this in the research on gender. In the 1970s, the research focus was on sex differences, particularly in achievement and participation in science at school and beyond, and the explanations for those differences reflected a "deficit" perspective. These explanations regarded the level of participation of white males as the norm, the nature of science as objective and immutable, and sought reasons for females' lack of participation in terms of their difference when compared to males. Concurrent strategies aimed at decreasing differences relied on compensating for the skills (such as spatial ability) and experiences (such as involvement in science-related activities out of school) that females were perceived to lack and which were believed to act as barriers to their participation in science.

In the 1980's, research in science classrooms identified aspects of pedagogy and practice that were sexist and based on stereotypes, such as the patterns of teacher-student interactions and the images of gender and science presented in curriculum materials and textbooks. Gender researchers' efforts were directed toward finding curricular and pedagogical strategies that recognized and accommodated the intellectual, physical, and emotional needs, learning styles, and values of both girls and boys in science classrooms. At this time, researchers also investigated issues about race, class, and locality which acted to preclude the participation of non-whites in science and the scientific community (e.g., Collins & Matyas, 1985; Malcom, Hall, & Brown, 1976; NSF, 1986). Considerable funds were directed toward the development of science materials and resources as well as teaching strategies that were variously labeled as "girl-friendly", "non-sexist", or "sex-equitable". Over time, researchers began to attend more carefully to the ways people constructed their own views of science and science learning. By listening to girls and boys, and to male and female teachers and scientists talking about their lives and their experiences in science, it became clear that if the issue of gender is to be considered meaningfully, account must be taken of the ways both gender and science are constructed in terms of social categories such as ethnicity, class, religion, and race, which in themselves are indistinct and intersect in multiple ways. Further, all of these categories need to be considered in terms of the prevailing socio-cultural context.

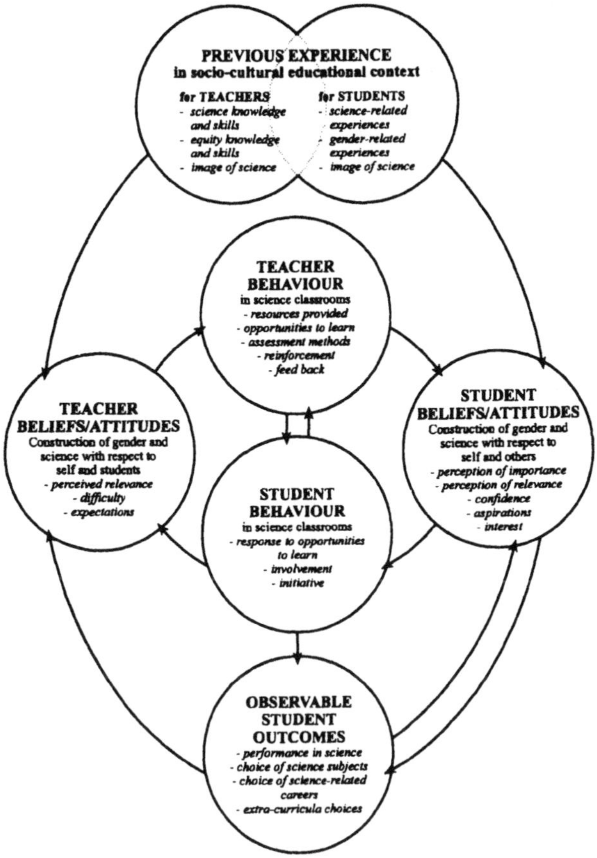

Figure 1: A socio-cultural model for explaining gender differences in science-related attitudes, perceptions, classroom behavior and learning outcomes (from Rennie, Parker, & Kahle, 1996, p. 215)

Thus, the frameworks used for conceptualizing gender equity in science and science education have changed. Initially, females were viewed as deficient in some way because they did not seem to suit science as well as males did. Subsequently, gender perspectives were more cognizant of science in its socio-cultural context. The perspective currently favored by gender researchers is a socially critical one, which questions how science is used in schools and in society to privilege some groups of people over others. Writing in mathematics education, Willis (1996) has described how this socially critical perspective views the curriculum as actively implicated in producing and reproducing inequality within many social categories, because, "through its [mathematics] content and practices, it constructs the successful mathematics learner as middle class and male" (p. 46). The parallel with Western science is obvious. It is the nature of science and the science curriculum, and the socially constructed practices which pervade them which need to be challenged, not the suitability of females (and males) from any socio-cultural background to be involved.

This brief summary of the changing directions in gender research emphasizes the point made earlier, that different perspectives for interpreting gender differences lead to different strategies for addressing those differences. All of those involved in the preparation of science teachers—scientists, science teacher educators, cooperating teachers in schools, prospective teachers themselves and pre-college students—construct their own perspectives about gender and science. These perspectives are based on the interaction between their knowledge of science and science education, and their gender-related experiences and observations in everyday life, as well as what happens in science classrooms. Because people's experiences are different, they are likely to hold different perspectives about gender and its relationship to science. Thus effective use of the model presented in Figure 1 requires the recognition that different people have different perspectives about gender equity, and that their actions and behavior are consistent with their perceptions and beliefs about gender and science and the relationship between them.

FOUR PERSPECTIVES FOR INTERPRETING THE MODEL

Gender-inclusive practice in science teacher preparation requires recognition of participants' initial perspectives about science and gender, and the use of strategies consistent with those perspectives to promote gender equity. Willis (1996) describes four perspectives about the mathematics curriculum, disadvantage and social justice that are most effective for the purpose of this chapter. Although developed and described in mathematics education, Willis' framework is equally applicable in science education, and its use can contribute to better communication in science methods classes and among science educators and researchers (Rennie, 1998). In the following sections, Willis' four perspectives are rephrased in terms of gender differences and science education.

A Remedial Perspective

From this first perspective, explanations for gender differences focus on the students and accept the science curriculum as entirely appropriate. If some students are disadvantaged in some way, the problem is considered to lie with them. Some students, because of the social group to which they belong (in this case, girls), are thought to be less well prepared than others (in this case, boys) to benefit from the science education they are offered. Viewed from the remedial perspective, the solution is compensatory, and lies not with changing the prevailing science curriculum, but with providing those disadvantaged students with the missing skills, experiences, or motivation they need to study science. This perspective is remedial, but it is widely known as a deficit model—"there's something wrong with/missing from girls, so we need to compensate them for it". For example, people with this perspective of gender difference might consider boys and girls to be equally good at science, but think that the girls lack interest and cannot see the benefits of studying science. Achieving gender equity from this perspective would mean encouraging the girls to take science by pointing out the benefits to them, but would not require changing the curriculum in any way.

A Non-Discriminatory Perspective

People who think about gender differences from this perspective focus on the way the curriculum is delivered. The problem of disadvantage is considered to lie in the way that the science curriculum is taught or assessed, although its content is regarded as unrelated to any disadvantage. Thus, if pedagogical practice or the way science is assessed favors the social and cultural background experiences of one sex more than the other, then both participation and outcomes will be gender-biased. For example, perhaps teachers spend more time interacting with boys than girls, perhaps the language, examples, and resources they use to explain concepts are more suited to boys' experiences, or perhaps the assessment tasks enable one group to demonstrate their knowledge and skills more easily than another group. From this perspective, the solution to the problem of gender difference is to take students' background and experiences into account and provide the kinds of classroom environment, learning activities, and assessment tasks that enable them to achieve their best. Increasing gender equity requires providing an equally supportive learning environment for girls and boys and equal opportunities for them to demonstrate what they have learned. Thus equity might require that students learn in different ways or are assessed by different tasks. The task for educators would be to eliminate those aspects of pedagogy and practice that are discriminatory and to employ non-sexist classroom strategies. Willis (1996) calls this a non-discriminatory perspective, or non-sexist when the focus is on gender.

An Inclusive Perspective

Explanations of gender differences from the third perspective challenge the science curriculum itself as the likely source of disadvantage. The curriculum is not regarded as fixed, but a selection from many different possible curricula. However, when its content and sequence reflect the kinds of dominant cultural and social values which are stereotyped with respect to gender, then students in non-dominant social groups, like females, are forced to learn a science which is less well matched to their interests and experiences. This happens even when the best pedagogical and assessment practices are used. Thus, for example, the science curriculum might reflect cultural values which privilege some characteristics, such as objectivity and rationality, over others, such as subjectivity and intuition. The curriculum might privilege those who do physics over those who do biology, for example, or value interpretations of natural phenomena based on reductionist, rather than holistic, explanations, instead of valuing the content which is most suitable for students' needs and interests. In terms of this perspective of disadvantage, the problem is tied up with the choices we make about the science curriculum, and the solution requires rethinking who the curriculum is for and how it can be improved so that it can work equally well for all students. Thus equity from this perspective requires thinking about the nature of all the students for whom the science curriculum is developed, and restructuring it to accommodate in a more inclusive way the interests, attitudes, social experiences, and values of all those students. When the focus is on gender, this perspective is usually referred to as gender-inclusive.

A Socially Critical Perspective

People who think about gender differences from this fourth perspective view the science curriculum as actively implicated in producing and reproducing gender inequality. Whereas the inclusive perspective views females as not included in science, this perspective views them as actively excluded. The content and practice of science in schools and society are seen to work to maintain the dominant culture, values, and group interests, as suggested by a view of science as male, white, Western, and middle class, and thus excludes others. When viewed from this perspective, the problem of disadvantage in science education can be interpreted in terms of how science is used both inside and outside of schools to position and privilege some people over others in ways which are based on gender, race, class, culture, locality and personal abilities. Thus science itself and the science curriculum are perceived as biased in a way which favors one group over another, in the case of gender, males over females. Willis teased out the educational consequences of adopting this socially critical perspective. She points out that the solution to the problem of inequity in this fourth perspective requires teachers and educators to challenge the hegemony of science in that particular socio-cultural context and modify its use to serve students in a way which is more fair and just in a social sense. This means that participants must first recognize the hegemony. From this perspective, our task as educators is to examine the ways in which science and the science curriculum are constructed and to reconstruct our views of who does science and what it means to be good at science. We can adapt Willis' (1996) words about mathematics to the context of this chapter. Thus, the aim for science teacher educators from the socially critical perspective is to assist prospective teachers, and through their efforts, students, "to understand how they and others are positioned by school [science] and to decide what they want to do about it, and how to use [science] in their own interests and in the interests of social justice" (p. 48).

MAKING USE OF THE MODEL AND THE FOUR PERSPECTIVES AS A FRAMEWORK FOR EQUITY IN SCIENCE TEACHER EDUCATION

Willis (1996) presented her four perspectives as a framework for others to use to "understand, compare and evaluate various strategies for addressing gender differences in school mathematics" (p. 51). She had found that locating the views of the particular educators with whom she was working within her broad perspectives "proved helpful in understanding their practices, their concerns, and their personal conflicts" (p. 50). These perspectives can offer the same approach for science teacher educators to understand the ideas and actions of prospective teachers (and indeed each other!) and to promote communication through a shared language. The perspectives are neither categorical nor mutually exclusive, and people's views may overlap, or move between, perspectives, depending on the issues involved. With this understanding, they present an effective place to begin.

Figure 2 presents a summary of the four perspectives as an overview of the framework as a model for action. It describes, from each of the four perspectives, assumptions about the cause of the disadvantage experienced by females, the nature of the science curriculum, the meaning of gender equity, the actions required to achieve it, and examples of strategies consistent with that perspective. In the context of the Kahle et al. model presented earlier in Figure 1, these four perspectives accentuate different clusters of variables. The remedial perspective attributes gender differences in observable student outcomes to students' beliefs and attitudes, either directly or indirectly though student behavior. If girls do not choose science, for example, an explanation from this perspective might be that they aren't interested or do not see science as relevant. The remedy would be to change girls' attitudes. The non-discriminatory perspective broadens the focus from student variables in the model to include teacher beliefs and attitudes and teacher behavior in the classroom, which relates to how the curriculum is delivered. If girls do not choose science, then an explanation might be that boys getting more of the teacher's time, which might unintentionally discriminate against girls in the classroom. A strategy toward gender equity from the non-discriminatory perspective would be to ensure that girls get an equal share of the teacher's time.

The inclusive and socially critical perspectives are broader still. They include all of the clusters of variables in the model, recognizing the role of the socio-cultural context in determining both teachers' and students' attitudes and beliefs, as well as the content of the curriculum and how it is enacted in the classroom. The difference is that the socially critical perspective does more than change the curriculum content and pedagogy to be more inclusive of the attitudes and experiences of all students, it actively challenges the exclusion from science of girls and other marginalized groups. The strategies for gender equity include teaching students about how our society positions males and females in science, and what might be done to challenge this.

If the equity principle of the *National Science Education Standards* and similar documents in other countries is to be achieved, it seems clear that science education must move toward the socially critical perspective. This perspective—the most conceptually demanding and least likely held by prospective teachers—offers the greatest potential benefit for change. Willis (1998) points out that this perspective

> does not demand of teachers, schools and educators generally that they either 'get the curriculum right' or they 'feel guilty'. Nor does it demand of girls and boys that they either become 'right minded' or 'feel guilty'. What it has the potential to do is provide them with some means for understanding how things are and how they came to be that way, and to understand the contradictions in their (and in our) feelings and actions. Thus, although the fourth perspective makes the most radical demands in terms of curriculum thinking, it is also the most supportive of the participants in the change process. (p. 17)

Assumptions:	*Remedial Perspective*	*Non-Discriminatory Perspective*	*Inclusive Perspective*	*Socially Critical Perspective*
The females from this perspective...	have a deficit which needs remediation	have been discriminated against	have not been included in science	have been excluded from science
The science curriculum from this perspective...	is unrelated to any disadvantage	has content unrelated to disadvantage, but the pedagogy and assessment are discriminatory	has content and pedagogy which are the cause of disadvantage (albeit unintentionally)	constructs disadvantage to maintain the status quo
Gender equity from this perspective is interpreted as...	equal (the same) provision and the same assessment tasks and contexts	equal opportunity to access the content and to demonstrate learning by equally fair assessment	equally appropriate curriculum	equality of outcome by group
Gender equity from this perspective requires...	compensatory education — along-side or prior to 'the curriculum' which is left unchanged	non-sexist learning opportunities and assessment practices	construction/reconstruction of the science curriculum to be more gender-inclusive	anti-sexist education, which teaches students about positioning and provides skills for social action
Example strategies include...	giving females help in "weak" areas, e.g. tinkering skills, confidence-building in doing science and being involved in science	monitoring equal access to teacher time, equal sharing of lab. work, etc., using non-sexist language and resources	variety in teaching and assessment strategies to value and respect diversity among students/prospective teachers	discussion of sexism, racism, science as a social construction, questioning how men and women are portrayed in science

Figure 2: Four perspectives of gender equity and the assumptions that underlie them (adapted from Willis, 1999, with permission)

In summary then, this chapter is based on a socio-cultural model (see Figure 1) that sets out clusters of teacher and student variables important in explaining gender differences. The model can be interpreted from four perspectives (see Figure 2), the remedial, non-discriminatory, inclusive and socially critical perspectives, each suggesting different kinds of strategies for ameliorating gender differences. I argue that the fourth perspective has the best fit with the equity principle of the *Standards*. Using the model to promote gender equity in science teacher education means recognizing and respecting the perspectives that participants in science teacher education have already constructed about gender and science, based on their own gender-related experiences in their personal, socio-cultural world. Importantly, it means recognizing that, as people's experiences accumulate and are reflected upon, their perspectives change. Enhancing gender equity requires identifying people's current perspectives and using appropriate strategies to scaffold that change toward a more inclusive, and a more socially critical perspective. Clearly, a range of strategies that can identify and accommodate a variety of viewpoints will need to become part of the science teacher educator's approach to gender equity.

STRATEGIES TO PROMOTE GENDER EQUITY

If prospective teachers do not possess particular skills or knowledge that are deemed prerequisite for effective teaching, they need to have opportunities to acquire those skills and knowledge. Apart from remedial action of this kind, remediation in general terms, especially on a gender basis, is a strategy that has limited value, and much less potential benefit than the other three perspectives in Willis' framework. There are numerous sets of guidelines to promote a more gender equitable science education, many of them aimed at the professional development of teachers. Examples include Ditchfield and Scott (1987), Gianello (1988), Hildebrand (1989), Lewis and Davies (1988), Mason (1995), Murphy and Gipps (1996), Treagust and Rennie (1989) and Weinburgh (1995). Although not aimed at science, Allard, Cooper, Hildebrand, and Wealands (1995) present a particularly comprehensive set of professional development materials for addressing gender in educational settings. There is not space to reiterate all these strategies here. The purpose of this chapter is served better by providing some examples of how science teacher educators can use the framework of the model and the four perspectives to work toward a more gender equitable science education.

Examples demonstrating two approaches are provided. One approach uses a critical incident relating to gender and science to provoke discussion with prospective teachers. Examples 1 and 2 are of this kind. By analyzing the gender issues, prospective teachers can explore and articulate their own views, reflect on them, and perhaps reconstruct their perspective. The second approach recounts the experiences of science teacher educators attempting to change their own teaching practice toward a more gender equitable curriculum. Examples 3, 4, and 5 describe the strategies used by these educators and documents their evaluation of their success.

Examples Using Critical Incidents

The first two examples are based on interview excerpts from a large study of receptivity to gender reform in Australian schools (Kenway & Willis, 1998). The excerpt gives clues to the perspective of the participants. Using the Kahle et al. model, prospective teachers can suggest strategies to move toward a more equitable position.

Example 1: Girls' Performance on Multiple Choice Tests

John (a pseudonym), a male science teacher with 25 years experience, is discussing the performance of 11th grade science students on an Australia-wide Chemistry Competition with researchers Léonie and Maggie.

John: Yeah, this year the boys are pretty strong. They are successful. What is very visible is that we are giving them multiple choice and they did the Chem Exam [for the competition] in the lesson during the Science in Schools Week and the boys actually left them [the girls] for dead. The girls are definitely disadvantaged.

Léonie: Do they do well in other exams?

John: It's the multiple choice. I discussed that with the 11th grade and they think it is unfair.... Because invariably girls do poorly... Well, we had the chem one [the chemistry competition test]. They all did it, we made them, well, we just gave them all out one [copy of the test].

Léonie: Yes?

John: And the guys got about seven or eight certificates [commendations] and the girls got one or two, I think. (Laughter) So they weren't very successful with that and that was just straight multiple choice. That is one thing I do feel, that maybe something odd happened. (Pause) I think the girls must read them and they see the right answer and they think maybe it is this one or maybe it is that one and maybe they can sort of see the value in lots of answers and then they just go for all. I am sure (pause) the way I do it is that I find that is the right answer and I just find that, I don't even look at the rest. Just go for the first one you think is right. Maybe the boys gamble a bit more. I don't know what it is.

Maggie: How do you feel about the notion that girls are more inclined to be communicative in their answers so that they might like to write something around the answer...

John: Right!

Maggie: and when they are constrained to select?

John: I think that they are, yeah, what is the right word? I think that they are not prepared to take a risk, I think. They are not confident enough to say, "Well that is the answer". They say, "Oh," and because they are proper constructed questions they are obviously taking the answers on their merit and they just get nervous about making a decision. Whereas the boys say, "Oh, that is the right answer. Next one!" If they are smart enough to be spontaneous then they can get things all right. They have a gut feeling I suppose, whereas the girls tend to go, "Oh well, there is too much information. I don't know". I have spoken to a few of them in 11th grade about that. Like I was saying, they think it is unfair that it happens to be multiple choice. I told them that it was 30% in the TEE [tertiary entrance examination] that way so they had better get some practice in. (Laughter)

Participant's perspective. John's comments suggest that he blames girls' poor performance on the chemistry test on their lack of confidence and risk-taking in multiple choice tests. He is somewhat perplexed by their approach—it is not the way he or the boys do these questions—and he sees it as a potential problem for them because later examinations will include multiple choice items, so the girls "had better get some practice in". From this limited evidence, it appears that John's perspective on this gender difference is best described as remedial—the girls have a problem and they need extra practice to compensate for it.

Discussing strategies. Discussion of an incident such as this with prospective teachers could explore other interpretations of the difference and other solutions. For example, using the Kahle et al. (1993) model, John's view is that the girls' attitude (unwillingness to take risks) is responsible for their ineffective test-taking behavior and subsequent poor performance. But the model also suggests that John's own behavior and particularly the assessment methods he uses in class could have some effect on the issue. It is easy to see how a non-discriminatory or even a gender-inclusive solution to the problem could be suggested in discussion. Some prospective teachers may want to trace girls' attitudes back to the socio-cultural coeducational context and question why it endorses the stereotype of girls as low risk-takers compared to boys. This begins a move toward a socially critical perspective.

Example 2. Gender-Based Harassment

Jan (a pseudonym) is an experienced female English as a Second Language (ESL) teacher in a high school where more than 40 different languages are spoken by students. The excerpt is from a discussion about gender issues and how they are perceived within the school.

> Léonie: Is gender an issue which interacts in any way with the ESL nature of the school?
>
> Jan:A lot of girls tend to be more obviously passive, submissive, less domineering because they come from gender traditional backgrounds. I suppose it relates in that way. We did have a situation where a boy and girl came from the same ethnic background. Although they were unrelated, the guy was constantly critical of the girl because she wasn't behaving in a traditional way. Her behavior was perfectly acceptable for Australian society in this school, but it wasn't acceptable to him, because he didn't think she should wear shorts to school, or voice an opinion.

Participant's perspective. This example is not related specifically to science, but this sort of thing happens in science classrooms. In the report of the large project, Kenway and Willis (1998) give numerous examples of the intersection between gender and ethnicity and culture, and they are the kinds of issues which teachers need to deal with, both in and out of the classroom. Jan was describing an incident of gender-based harassment, but it occurred only because both students were from the same ethnic group. Jan's perspective is a socially critical one, and she explains the students' behavior in terms of the gendered expectations of their culture. The boy believed that appropriate behavior for a girl did not include wearing shorts or expressing her own opinion, and he felt he had the right to criticize. However, in the culture of this school, the girl's behavior was not only acceptable, but expected if she was to fit in with the other girls. Jan's intervention was a discussion with the boy to explain this alternative perspective and stop the criticism from escalating, and she felt it was successful.

Discussing strategies. In the Kahle et al. model, the incident draws attention to differences in students' attitudes and beliefs in a socio-cultural context and to the way this affects their interactions with each other. When conflict arises, teachers will call upon their own beliefs and values to choose how they will act in ways which can resolve the conflict. Was Jan's intervention the "right" thing? Is there a "right" thing in issues such as these? Discussion of incidents like this one provide opportunities to look beyond gender as the only explanation for behavior among boys and girls and to look instead at the students as individuals. As Rizvi (1995) points out, "no category of social difference is homogeneous" (p. 28). In this instance, gender was important but the problem was the conflict between what was appropriate behavior from an ethnic perspective, and what was appropriate behavior in the culture of the school.

Examples of Changing Pedagogy in Science Teacher Education

If we aim to prepare science teachers who meet the *National Science Teaching Standards* for equity in science teaching, how we teach is a significant factor. The examples just described are aimed at giving prospective teachers opportunity, in a non-threatening way, to challenge and reconstruct their perspectives about gender and science. A stronger approach is to build this challenge into science methods courses by shaping the content and pedagogy in ways that model a socially critical perspective. The next three examples outline the attempts of science teacher educators to restructure their science methods courses in ways that they hoped would allow prospective teachers to challenge their views of science, to feel more connected with it, and to re-envision it in ways which were more inclusive and empowering.

Example 3. A Feminist-Constructivist Approach

Roychoudhury, Tippins, and Nichols (1995) drew implications from feminist theories which they felt were fundamental to addressing gender issues in science education. They described these as "(a) situating science learning in the lived experiences of students; (b) assigning longer projects to allow a development of personal bonding with learning experiences; and (c) providing a cooperative and supportive environment" (p. 902). These authors recognized that the curriculum would require a good deal of flexibility to accommodate different learning preferences, interests and assessment methods. As well, they believed that all of the learners, both male and female, needed to experience different ways of knowing.

The authors chose a physical science course for prospective elementary teachers to try out their ideas. Three principles were developed to put the feminist recommendations into practice, particularly for the open-ended projects which were part of the course. The principles were to "(a) encourage personal experience and interests of students, (b) increase the number of observations and make the projects at least a few weeks long so that students develop a connection with the subject of their study, and (c) use less competitive practice by encouraging group work as much as possible" (Roychoudhury et al., 1995, p. 905). Overall, the course was successful with Roychoudhury et al. reporting that 70% of the 45 prospective teachers were very positive and 15% were cooperative but less enthusiastic, making only brief reflections and not appearing excited about the activities or projects. A few others disliked the course or parts of it claiming that the cognitive demand was too high, they didn't receive enough information about what to do, or they didn't know what the "correct answer" was meant to be.

Roychoudhury et al. report that over 80% of the class were initially uncomfortable with science and, while the evidence suggested great improvement, many prospective teachers were frustrated dealing with the uncertainty of data. Roychoudhury et al. emphasize the importance of developing a sense of ownership and empowerment in the learning of science, particularly for the females they described as having "become strangers to science" (p. 917).

Example 4. Challenging Science, Challenging Gender

Haggerty (1995, 1996) reports on aspects of a long term action research project with prospective high school teachers enrolled in a science methods course. One of the goals was to challenge prospective teachers' ideas about the nature of science, the role of the science teacher, and the nature of learning about science, leading them to become more aware of their practice and its implications for their students. About equal numbers of males and females were involved, all with some tertiary background in science. Haggerty (1996) reports that most of the prospective teachers initially demonstrated an empiricist/positivist view of science. Throughout, the course emphasized a social constructivist view of learning and of science, and readings, discussions and feedback, and teaching strategies explicitly encouraged reflection on their practice.

By the end of the one-year course, some changes were evident but the prospective teachers still struggled with the idea that science and scientific knowledge are socially constructed and that its tenets are consensual rather than realist. For the most part they held naive views about the role of theory in science. From data collected in another part of the project, Haggerty (1995) reports the mixed responses prospective teachers gave to some of the gender-related aspects of the course. Most of the participants were unconcerned about gender issues, but the reactions of others included two who made explicit attempts in practice teaching to encourage girls in science, and some (including males and females) who were angry at the time spent on the topic considering it to be wasted. Overall, students were much more willing to accept a constructivist view of students' learning than they were to accept a constructivist view of science and scientific knowledge.

Example 5. A Feminist Pedagogy

Richmond, Howes, Kurth, and Hazelwood (1998) present four separate autobiographical case studies of science teacher educators striving to implement a feminist pedagogy in their teaching. They each used three themes to characterize their pedagogical approaches and to structure their reports. These were, "helping our students *rethink* their connections with science, Helping our students to *re-envision* science, helping our students to *transform* these perspectives into a pedagogy they own and that attracts, enlightens, and empowers the students they teach" (p. 900, original emphasis). Richmond et al. (1998) based their reports on analyses of assignments designed to demonstrate prospective teachers' progress in rethinking, re-envisioning and transforming their perspectives. All document some progress but also some difficulties. The courses included a mixture of science and science methods for prospective elementary and secondary teachers, so perhaps differences might be expected. One, for example, was that prospective elementary teachers were much less connected to science than those who were planning to be secondary teachers, and it was threatening for many of this latter group to critique science when they felt some authority from their experiences within it. Some felt, as had several of the participants in Haggerty's study, that time spent on critique was time that could have been spent in learning how to teach. The authors (Richmond et al., 1998) conclude:

> to teach effectively is not to destroy the shape of scientific practice as it existed in the past, but to recognize the factors that have created its shape and scope, as well as its limitations, to understand how it can be reshaped so that it invites in multiple perspectives and diverse groups of practitioners, and to reshape it along with one's students, who, in the process, become empowered to critique, extract, and then, perhaps most important, go on to design their own ways of engaging in scientific investigation. (p. 916)

Discussion of Examples 3, 4 and 5

All three of these examples demonstrate the kind of socially critical perspective that has been advocated in this chapter. Although each group of authors has used a different term, each describes challenges to the way science is constructed and, perhaps most explicitly in Haggerty's (1995) paper, how it is gendered. All three examples emphasize empowerment as a critical issue, aiming to give prospective teachers the skills, confidence, and ability to do science for themselves, to become part of its culture rather than be outsiders. But for some secondary prospective teachers, who already had a background in science, such critique was difficult and threatening. Perhaps more success was achieved overall with elementary prospective teachers (mostly women) who, more commonly, felt initial disconnection from science. Their qualified success shows that introducing a socially critical perspective in science teacher preparation is not an easy task.

Other researchers have found that dealing overtly with issues like gender equity is likely to be sensitive (Kenway & Willis, 1998; McGinnis & Pearsall, 1998) and this sensitivity can be traced back to the different experiences and perspectives of the prospective teachers, school personnel, and others. It is not only gender issues which can provoke resistance, of course. Rodriguez (1998) describes two types of resistance from prospective science teachers to multicultural education: resistance to ideological change and resistance to pedagogical change. There is a parallel with gender, with a resistance to adopting a socially critical perspective, particularly by those holding a remedial perspective, and resistance to implementing the kind of strategies which such a perspective would imply. The likelihood of resistance makes it especially difficult, but all the more important, for science teacher educators to recognize, acknowledge and value the experiences of the participants in respectful and involving ways. Potentially, this offers the best approach for science teacher educators to provide prospective teachers with the skills needed to establish a gender equitable science education.

SUMMARY AND CONCLUDING COMMENTS

This chapter is based on the assumption that "all students should learn science through full participation and that all are capable of making meaningful contribution in science classes. The nature of the community in which students learn science is critical to making this assumption a reality" (NRC, 1996, p. 46). The chapter stresses that all students are included only when they are treated according to their needs as individuals, not according to their membership of a social group based on gender or culture. Students in science classrooms, prospective teachers in science methods classes, and science teacher educators themselves, have had different experiences and hold different perspectives about gender and science. A gender equitable science education is possible only when the knowledge and understanding of all participants are recognized and valued.

The chapter provides a framework for thinking about the relationship between gender and science which can be used to suggest suitable strategies to promote gender equity in science teacher education. The framework is based on a socio-cultural model (Kahle et al., 1993; Rennie et al. 1996) presented in Figure 1, which maps clusters of variables associated with thinking about gender differences. Four perspectives (Willis, 1996) are used to interpret the model and to explain these differences. Each perspective accentuates different parts of the model and, accordingly, each perspective leads to different kinds of strategies to ameliorate gender differences (see Figure 2).

The socially critical perspective is the one most likely to support change, because it does not attribute blame or guilt about gender differences. Instead it provides a means to understand and challenge the social stereotypes about gender and science that permeate the ways we think and behave, and to suggest what steps might be taken to achieve the equitable science education described in the *Standards*. Unfortunately, it does not mean that the steps are easy to take! In the next sections, three principles are suggested to guide the use of the model and perspectives in science teacher education, followed by five criteria by which the success of their application may be judged.

Three Guiding Principles

Start where the prospective teachers are at. Any group of prospective teachers will vary in their content knowledge of science and have a variety of views about science and about gender, some of which may not be well-formed. Others may have very firm beliefs about science based on their cultural or religious beliefs. Expanding their perspectives requires time for prospective teachers to become aware that gender issues are important. They require time to articulate and evaluate their own ideas, and decide whether they want to challenge them. Science teacher educators need to understand what the prospective teachers are thinking so that scaffolding can be provided to help bridge the gap between one perspective and another. Examples 1 and 2 demonstrate how to use critical incidents to engage prospective teachers in discussion of gender issues at the level of the perspective they currently hold, enabling them to analyze and reconstruct their perspectives.

Be opportunistic. Teacher educators need to seize every opportunity to have prospective teachers examine their own beliefs and perspectives, but in ways which are cooperative and collaborative, rather than confronting. Such examination also provides opportunities to practice the skills of critical analysis. Real classroom incidents, in the methods class or during teaching practice, can serve as a basis for discussion and exploration of the relevant issues for gender and science using the strategies outlined in Examples 1 and 2. The first-hand experiences of participants (as they arise, for example, in the coteaching approach presented by Roth, this volume) have an immediacy and relevance that are difficult to match with constructed vignettes, although these and published research have a place.

Model gender-equitable pedagogy, assessment and inclusive curriculum. The notion of "do as I say, not as I do" has no place in equity issues because values, attitudes, and opinions cannot be mandated. Science teacher educators must lead by example. If we want their eventual classes to be gender equitable, then we must ensure that our prospective science teachers learn in a community which also is gender equitable. Examples 3, 4, and 5 outline the strategies used by some science teacher educators to move explicitly in this direction. They describe the strategies used by science teacher educators to make their own practice more socially critical, by encouraging and empowering prospective teachers to challenge and rethink the nature of science and their place in it.

How Can We Judge Success?

How can we tell whether using strategies in the framework of the model and the perspectives helps prospective teachers to develop their perspectives about gender and science? Is there movement toward a more gender equitable science education? It might sound counter-productive, but a little consternation among participants is probably a good sign, because it means they are thinking about the issues. More positively, success will be evident when prospective teachers

1. recognize that gender differences are not problems to do with girls, or with boys, but with a range of variables concerned with how science and gender are socially constructed;
2. acknowledge that students have different starting points and learning rates and take steps during lesson planning to provide them with a variety of ways to learn and to demonstrate their learning;
3. recognize that equity is an integral part of good teaching and learning, therefore time spent on it in science methods classes is not wasted;
4. are comfortable with, and feel connected to, science as a socially constructed way of thinking about the world; and
5. are comfortable discussing gender and gender equity and actively search for ways to make their science teaching more inclusive of all student groups.

A Final Note

The underlying premise of most strategies designed to promote gender equity is that opportunities for observation, reflection, and supportive discussion about gender-related issues in science teaching will assist prospective teachers to challenge their current attitudes and beliefs, and to shape them in ways which are more consistent with a socially critical perspective. Essentially, such strategies will need to encourage collaboration and cooperation among all participants. In practice, few strategies can be implemented without some resistance. Attempts to challenge the status quo from a gender-base invariably raise emotion. For example, an intervention which tries to promote equal opportunity for girls to interact in the classroom can be viewed by participants, both female and male, as an attempt to favor girls over boys and therefore inequitable. Pointing out that girls do not receive as many opportunities to interact as boys and that the intervention is attempting to make things more equal (the non-sexist perspective) seems to blame the boys, and both boys and girls resist the inference of blame. Instead, if the gender lines were not drawn, and the intervention was aimed explicitly at getting more students involved in classroom interaction (most of whom might be girls, but some might be quiet boys), less resistance might be encountered, because students are being treated according to their individual needs, and their sex is not relevant.

Kenway and Willis (1998) discuss the outcomes of their extensive Australian study of receptivity to gender reform and make this point:

> Anyone who has attempted gender reform in schools knows that it is much easier said than done. Gender is deeply and often unconsciously ingrained within people's psyches and behavior and deeply inscribed within school cultures and education systems. Therefore, changing girls, boys, teachers and school curricula and cultures is no simple matter... it provokes complex and often unexpected reactions. Gender reform in schools is a process fraught with difficulty. (p. xiii)

Efforts for reform in science teacher education are no less fraught. After all, prospective teachers expect their training to teach them about teaching, and if they don't consider gender to be a relevant issue, then they are liable to regard time spent on gender issues as wasted. Perhaps only small changes in prospective teachers' perspectives might be hoped for in a teacher-training program. Nevertheless, we must work at it if we are to move toward achieving the goal of science for all.

REFERENCES

Allard, A., Cooper, M., Hildebrand, G., & Wealands, E. (1995). *STAGES–Steps towards addressing gender in educational settings.* Carlton, Victoria: Curriculum Corporation.

American Association of University Women. (1992). *How schools short change girls.* Washington, DC: Author.

Baker, D. R. (1998). Equity issues in science education. In B. J. Fraser & K. Tobin (Eds.), *International handbook of science education* (pp. 869-895). Dordrecht, Netherlands: Kluwer Academic Publishers.

Baker, D., & Scantlebury, K. (Eds.). (1995). *Science "coeducation": Viewpoints from gender, race and ethnic perspectives.* Manhattan, KS: National Association for Research in Science Teaching.

Becker, B. J. (1989). Gender and science achievement: A reanalysis of studies from two meta-analyses. *Journal of Research in Science Teaching, 26,* 141-169.

Black, P. (1995). Editor's introduction. *Studies in Science Education, 26,* 1-6.

Catsambis, S. (1995). Gender, race and ethnicity, and science education in the middle grades. *Journal of Research in Science Teaching, 32,* 243-257.

Collins, A. (1995). National Science Education Standards in the United States: A process and a product. *Studies in Science Education, 26,* 7-37.

Collins, M., & Matyas, M. L. (1985). Minority women: Conquering both sexism and racism. In J. B. Kahle (Ed.), *Women in science: A report from the field* (pp. 102-123). Philadelphia: Falmer.

Curriculum Council. (1998). *Western Australian curriculum framework.* Perth, Western Australia: Author.

Ditchfield, C., & Scott, L. (1987). *Better science: For both boys and girls* (Secondary Science Curriculum Review, Curriculum Guide 6). London: Heinemann Educational Books/Association for Science Education for the School Curriculum Development Committee.

Fennema, E. (1987). Sex-related differences in education: Myths, realities, and interventions. In V. Richardson-Koehler (Ed.), *Educators' handbook: A research perspective* (pp. 329-347). New York: Longman.

Gianello, L. (1988). *Getting into gear—Gender-inclusive teaching strategies.* Canberra, Australia: Curriculum Development Centre.

Haggerty, S. M. (1995). Gender and teacher development: Issues of power and culture. *International Journal of Science Education, 17,* 1-15.

Haggerty, S. M. (1996). Towards a gender-inclusive science in schools: Confronting student teachers' perceptions and attitudes. In L. H. Parker, L. J. Rennie, & B. J. Fraser (Eds.), *Gender, mathematics and science: Shortening the shadow* (pp. 17-27). Dordrecht, Netherlands: Kluwer Academic Publishers.

Hildebrand, G. (1989). Creating a gender-inclusive science education. *The Australian Science Teachers' Journal, 35*(3), 7-16.

Kahle, J. B. (Ed.). (1985). *Women in science: A report from the field.* Philadelphia: Falmer.

Kahle, J. B. (1988). Gender and science education II. In P.J. Fensham (Ed.), *Development and dilemmas in science education* (pp. 249-265). London: Falmer.

Kahle, J. B., & Meece, J. (1994). Research on gender issues in the classroom. In D. L. Gabel (Ed.), *Handbook of research on science teaching and learning* (pp. 542-557). New York: Macmillan.

Kahle, J. B., Parker, L. H., Rennie, L. J., & Riley, D. (1993). Gender differences in science: Building a model. *Educational Psychologist, 38,* 379-404.

Kenway, J., & Willis, S. (with Blackmore, J., & Rennie, L.). (1998). *Answering back: Girls, boys and feminism in schools.* New York: Routledge.

Lee, V. E., & Burkam, D. T. (1996). Gender differences in middle grade science achievement: Subject domain, ability level, and course emphasis. *Science Education, 80,* 613-650.

Lewis, S., & Davies, A. (1988). *GAMAST professional development manual: Gender equity in mathematics and science.* Canberra, Australia: Curriculum Development Centre.

Linn, M.C., & Hyde, J.S. (1989). Gender, mathematics, and science. *Educational Researcher, 18*(8), 17-19, 22-27.

Malcom, S. M., Hall, P. Q., & Brown, J. W. (1976). *The double bind: The price of being a minority woman in science.* (AAAS Report No. 76-R-3). Washington, DC: American Association for the Advancement of Science.

Mason, C. (1995). Gender equity is still an issue: Refocusing the research agenda. In D. Baker & K. Scantlebury (Eds.), *Science "coeducation": Viewpoints from gender, race and ethnic perspectives* (pp. 7-21). Manhattan, KS: National Association for Research in Science Teaching.

McGinnis, J. R., & Pearsall, M. (1998). Teaching science methods to women: A male professor's experience from two perspectives. *Journal of Research in Science Teaching, 35,* 919-949.

Murphy, P. F., & Gipps, C. V. (Eds.). (1996). *Equity in the classroom: Towards effective pedagogy for girls and boys.* London: Falmer Press and UNESCO Publishing.

National Research Council. (1996). *National science education standards.* Washington, DC: National Academy Press.

National Science Foundation. (1986). *Women and minorities in science and engineering.* Washington, DC: Author.

Parker, L. H., Rennie, L. J., & Fraser, B. J. (Eds.). (1996). *Gender, science and mathematics: Shortening the shadow.* Dordrecht, The Netherlands: Kluwer Academic Publishers.

Parker, L. H., Rennie, L. J., & Harding, J. (1995). Gender equity. In B. J. Fraser & H. E. Walberg (Eds.), *Improving science education* (pp. 186-210). New York: National Society for the Study of Education.

Reid, A. (1995). The National Curriculum: Mechanism of control or a liberating curriculum tool? In M. Kalantzis (Ed.), *A fair go in education* (pp. 11-17). Canberra, Australia: Australian Curriculum Studies Association in association with the Australian Centre for Equity through Education.

Rennie, L. J. (1998). Gender equity: Toward clarification and a research direction for science teacher education. *Journal of Research in Science Teaching, 35,* 951-961.

Rennie, L. J., & Parker, L. H. (1993). Curriculum reform and choice of science: Consequences for balanced and equitable participation and achievement. *Journal of Research in Science Teaching, 30,* 1017-1028.

Rennie, L. J., Parker, L. H., & Kahle, J. B. (1996). Informing teaching and research in science education through gender equity initiatives. In L. H. Parker, L. J. Rennie, & B. J. Fraser (Eds.), *Gender, science and mathematics: Shortening the shadow* (pp. 203-221). Dordrecht, Netherlands: Kluwer Academic Publishers.

Richmond, G., Howes, E., Kurth, L., & Hazelwood, C. (1998) Connections and critique: Feminist pedagogy and science teacher education. *Journal of Research in Science Teaching, 35,* 897-918.

Rizvi, F. (1995). 'Broadbanding': Equity in Australian schools. In M. Kalantzis (Ed.), *A fair go in education* (pp. 25-31). Canberra, Australia: Australian Curriculum Studies Association in association with the Australian Centre for Equity through Education.

Rodriguez, A. J. (1998). Strategies for counterresistance: Toward sociotransformative constructivism and learning to teach science for diversity and for understanding. *Journal of Research in Science Teaching, 35,* 589-622.

Roychoudhury, A., Tippins, D. J., & Nichols, S. E. (1995). Gender-inclusive science teaching: A feminist-constructivist approach. *Journal of Research in Science Teaching, 32,* 897-924.

Shroyer, M. G., Smith, N. J., Borchers, C. A., & Wright, E. L. (1994). Science and mathematics equity issues at a local School District level. *School Science and Mathematics, 94,* 65-77.

Sjøberg, S., & Imsen, G. (1988). Gender and science education I. In P. J. Fensham (Ed.), *Development and dilemmas in science education* (pp. 218-248). London: Falmer.

Treagust, D. F. & Rennie, L. J. (Eds.). (1989). Gender-inclusive science and technology education [Special Issue]. *Australian Science Teachers Journal, 35*(3), 1-120.

Weinburgh, M. (1995). Preparing gender-inclusive science teachers: Suggestions from the literature. *Journal of Science Teacher Education, 6,* 102-107.

Willis, S. (1996). Gender justice and the mathematics curriculum: Four perspectives. In L. H. Parker, L. J. Rennie & B. J. Fraser (Eds.), *Gender, mathematics and science: Shortening the shadow* (pp. 41-51). Dordrecht, Netherlands: Kluwer Academic Press.

Willis, S. (1998). Perspectives on social justice, disadvantage, and the mathematics curriculum. In C. Keitel (Ed.), *Social justice and mathematics education: Gender, class, ethnicity, and the politics of schooling* (pp. 1-19). Berlin: International Organisation of Women and Mathematics Education (IOWME) and Freie Universität Berlin.

Willis, S. (1999). La igualdad entre los sexos y el plan de estudios de matemáticas en las escuelas: algunas consideraciones desde Australia (Gender equity and the school mathematics curriculum: some views from Australia), *UNO, 19,* 71-85.

ASSESSMENT MODELS THAT INTEGRATE THEORY AND BEST PRACTICE

Mary Stein
Wayne State University

INTRODUCTION

When it comes to issues of classroom assessment, science teacher educators must ask themselves, "Are we practicing what we are preaching?" Today, more than ever before, a wide range of assessment strategies is promoted for use in science classrooms (Berenson & Carter, 1995; Council of Chief State School Officers, 1997). As school districts are changing their science programs to more closely align with outcomes suggested by national, state, and local reform efforts, the need for a variety of assessment measures that target these outcomes has become apparent. With more problem solving, process skills, and inquiry oriented strategies being incorporated into the science curricula, teachers need to know how to accurately measure these elements of scientific literacy. This is not to say that students do not need content knowledge and factual information, but that scientific literacy involves a broad range of competencies (NRC, 1996). It is the emphasis on what students should know and be able to do that is changing, and as it does, assessment practices are also changing. Are our teacher education programs keeping pace with these changes? How many prospective science teachers have experienced alternative assessment strategies as an integral component of their teacher preparation program? And if they have not, can we really expect that new teachers will be able to implement alternative assessment strategies successfully?

There are a wide variety of assessment formats that can provide different kinds of information and that can be used for different purposes. The focus of this chapter is on the use of portfolios and performance assessments as integral components of science teacher preparation programs. Schools often look towards teacher education programs, colleges and universities, as well as the new teachers emerging from them, as a source of ideas and strategies that can move the profession forward. Several states have begun using portfolios as a regular part of their teacher assessment procedures (Lomask, Seroussi, Budzinski, 1997; Lyons, 1998; McLarty, Furtwengler, & Malo, 1985). Teaching portfolios are part of the assessment system for teachers who seek National Board Certification (Barringer, 1993; NBPTS, 1997). Officials of the National Council for the Accreditation of Teacher Education have also indicated that portfolios will become increasingly important as evidence of student learning and performance.

D.R. Lavoie and W.-M. Roth (eds.), Models of Science Teacher Preparation, 149–162.

The development of portfolios as an assessment tool aligns closely with a learner-centered philosophy (Paris & Ayers, 1994). Based on learner-centered psychological principles developed by the American Psychological Association, Paris and Ayers developed twelve learner-centered principles of assessment. The emphasis of these principles is on ongoing assessment, self-assessment, equity, and on the multi-faceted nature of assessment. With increased emphasis on portfolios and performance assessments in K-12 classrooms, teacher preparation programs must model the assessment strategies their students will be expected to implement in their own science classrooms.

THE CHANGING VIEW OF ASSESSMENT

Many researchers have called for a broadened and integrated view of assessment and instruction that better reflects a conceptual change or constructivist paradigm for science education (Angelo, 1995; Collins, 1992; Doran, Chan, & Tamir, 1998). Some have viewed assessment solely as a means for ranking and comparing students with each other or against some specific criteria. An enriched view of assessment and its purposes includes using ongoing assessment measures to inform instruction and provide a means for students to reflect on their beliefs. For example, Angelo (1995) focuses on assessment as a tool for improving student learning:

> Assessment is an ongoing process aimed at understanding and improving student learning. It involves making our expectations explicit and public; setting appropriate criteria and high standards for learning quality; systematically gathering, analyzing, and interpreting evidence to determine how well performance matches those expectations and standards; and using the resulting information to document, explain, and improve performance. When it is embedded effectively within larger institutional systems, assessment can help us focus our collective attention, examine our assumptions, and create a shared academic culture dedicated to assuring and improving the quality of higher education. (p. 8)

This definition has important implications for science teacher preparation programs. Thus, many teacher education programs and professional organizations are exploring and implementing alternative assessment measures, such as the use of portfolios, as a means of assessing teaching practice. As many have attested, the assessment of teaching practice is not simple, nor straightforward. Teaching is a complex activity and there is agreement that multiple sources of data are needed to capture this complexity inherent in teaching (Shulman, 1987; Wolf, 1991). Teaching is complex, multifaceted, and situated and involves observable as well as tacit knowledge (e.g., Roth, this volume). This implies that assessment of teaching must also be multifaceted, be informed by the context, and provide a means for articulating underlying beliefs about practice (Varvus & Collins, 1991).

PERFORMANCE ASSESSMENTS AND CONSTRUCTIVISM

Many inservice teachers are engaged in professional development experiences that help them to align assessment practices with instruction, as well as with their beliefs about teaching and learning. One study with middle-level science teachers indicated that beliefs about learners informed their assessment practices (Shepardson & Adams, 1996). Specifically, they recommended that preservice and inservice courses need to address assessment and pedagogical issues from the perspective of learners by integrating knowledge of learners, pedagogy, content, and assessment. These suggestions echo the recommendations of others who suggest that educators should move toward learner centered, or constructivist approaches, in the design of science programs and assessments. Approaches consistent with constructivist perspectives on learning focus on what students know or believe. As students interact with each other while exploring their beliefs in new and different contexts, students construct new and enhanced understandings.

Some of the underlying tenets of constructivism are evident during the process of developing and creating a portfolio. Creating a portfolio is more than simply a process of gathering samples of students' work. Consistent with a constructivist view of learning, portfolios involve students in actively constructing knowledge in the context of their experiences. Therefore, portfolios are a means to provide a more equitable and sensitive portrait of what students know and are able to do. A key aspect of the portfolio process is the reflection that must take place over time as students develop their portfolios (Reistetter & Fager, 1995).

PERFORMANCE ASSESSMENTS IN SCIENCE EDUCATION COURSES

The goal of science teacher educators is to prepare teachers to help their students become scientifically literate. Scientific literacy is complex and embodies a wide range of knowledge, skills, and dispositions. Similarly, the assessment of scientific literacy needs to be multifaceted and requires the use of a variety of assessment formats. Our science education students are likely to have a depth of experience with traditional paper-and-pencil assessment formats such as multiple-choice, true/false, short-answer, and essay examinations. The questions in these assessments typically tend to measure low-level cognitive skills rather than elements of scientific literacy such as problem solving, inquiry, communication, and interpersonal skills. How many teacher education students have experienced alternative assessment formats that focus in on these other aspects of student learning?

There are a variety of alternative assessment formats for science education that have been detailed elsewhere (e.g., Doran, Chan, & Tamir, 1998). Students can experience some of these formats, and experience first hand knowledge of their various strengths and weaknesses, through integrating various formats of assessment into existing science education courses. Design, development, and implementation of high-quality assessments are complex processes. Yet it is important that our students learn to assess in ways that are aligned with their beliefs about teaching. Some performance-based formats focus on problem-solving skills and process skills and are therefore based on (extended) investigations. Detailed examples of such assessment formats exist for biology, chemistry, earth science, and physics (Doran, Chan, & Tamir, 1998). The examples that they have outlined provide opportunities for middle level and high school students to show what they know and are able to do in ways that are not possible through traditional paper-and-pencil testing. For example, in a physical science course for elementary majors, students complete performance tasks that are similar to lab work as part of their examinations. This ensures that students are familiar with scientific equipment and processes before they begin working with younger students. The performance tasks mirror skills, processes, and activities that they have experienced previously during instruction, but in a new context. It is said that students do, and take seriously, what is included in assessments. Integration of performance tasks into the assessment system often motivates students to become more active in learning new skills during instruction when they know that these skills will be assessed later. At the conclusion of the course, students frequently comment on the value of experiencing and learning about new ways of integrating assessment with instruction.

THE CASE FOR PORTFOLIOS

The *National Science Education Standards* (NRC, 1996) refer to a view of assessment and learning as two sides of the same coin. The assessment standards included in the National Science Education Standards (NSES) include increased emphasis on using multiple assessment measures for a more complete understanding of student achievement:

> This view of assessment places greater confidence in the results of assessment procedures that sample an assortment of variables using diverse data-collection methods, rather than the more traditional sampling of one variable by a single method. Thus, all aspects of science achievement—ability to inquire, scientific understanding of the natural world, understanding of the nature and utility of science—are measured using multiple methods such as performances and portfolios, as well as conventional paper-and-pencil tests. (p. 76)

The National Science Education Assessment standards also highlight the importance of authentic assessment. This is described as assessments that require students to apply specific knowledge and reasoning to situations similar to those they will encounter in the world outside the classroom and situations that approximate how scientists do their work. The following assessment standards clearly articulate a need for changing assessments in a way that will better reflect a comprehensive view of scientific literacy.

1. Assessments must be consistent with the decisions they are designed to inform.
2. Achievement and opportunity to learn science must be assessed.
3. The technical quality of the data collected is well matched to the decisions and actions taken on the basis of their interpretation.
4. Assessment practices must be fair.
5. The inferences made from assessment about student achievement and opportunity to learn must be sound.

Other professional organizations involved with teacher education have also developed standards. A close look at these standards reveals that teacher educators have an important role in helping teachers understand effective assessment practices. The National Board for Professional Teaching Standards (NBPTS) has included assessment within Proposition 3, "Teachers are responsible for managing and monitoring student learning" (NBPTS, 1994). With respect to this proposition, the

NBPTS standards state that board-certified teachers should be able to assess the progress of individual students and employ multiple methods as part of the assessment process.

Similarly, the Interstate New Teacher Assessment and Support Consortium (INTASC, 1994) targets assessment in Standard 8, "The teacher understands and uses formal and informal assessment strategies to evaluate and ensure the continuous intellectual, social and physical development of the learner". It is clear that new teachers, as well as experienced professionals, are expected to implement multiple assessment strategies even when they may not have experienced these strategies as learners.

The Association for the Education of Teachers in Science (AETS) has developed a position statement on Professional Knowledge Standards for Science Teacher Educators. These standards for science educators are aligned with those for teachers. The AETS Standard 3.b states, "The beginning science teacher educator should possess expertise spanning a variety of assessment approaches, including 'traditional' and alternative assessment". It is clear that it is expected that teacher educators should have the knowledge and skills to help teachers understand and learn to implement a variety of assessment strategies.

One assessment strategy that is frequently listed in reform documents is the portfolio. The portfolio can provide a means for assessing what an individual knows and is able to do within the context of the learner's experiences. A portfolio can be defined as the collection, selection, and organization of teacher education students' work over time that shows evidence of self-reflection and learning (Wade & Yarbrough, 1996). Based on this simple description, it would be easy to overlook the tremendous impact portfolio development can have on a program. Portfolio development can change the climate of an educational program as well as the nature of student/faculty interactions. Well-designed portfolios represent important, contextualized learning and require complex thinking, reflection, and expressive skills. Not only can portfolios serve as assessment tools for individual students, they can also provide important indicators for course and program effectiveness.

Inherent in the portfolio assessment process is the constructivist notion that students have many different experiences, ideas, and approaches to learning. The portfolio process influenced positive changes in attitudes and beliefs concerning evaluation of teaching, professional growth, and reflective thinking, in pre-service teachers (Green & Smyser, 1995). Conversations occurring within the portfolio process engendered feelings of growth and confidence in preservice teachers (Freidus, 1998). For most students, the goal becomes not the final grade in a course or program but the creation of an accurate representation of what they know and are able to do in a way that best reflects their backgrounds and experiences.

As students collect, select, and present evidence of their understandings, the reflective component inherent in these processes generates an approach to learning in which instruction and assessment are integrated. When students create their portfolios with specific outcomes in mind, they think about the outcomes with respect to the meaning of the artifacts they have selected. They think about the quality and significance of their experiences and present the most relevant components in an organized manner in order to demonstrate professional competence.

A *portfolio culture* in science classrooms can be aligned with a constructivist orientation (Duschl & Gitomer, 1991). Thus, "the essential characteristic of this culture is that it creates opportunities for teachers and students to confront and develop their scientific understanding and to equip students with the tools necessary to take increased responsibility for their own restructuring, to assess for themselves what might be the next steps" (p. 840). Although much has been written on the development and use of portfolios, empirical research on this form of assessment is scarce (Herman & Winters, 1994). These authors found that of 89 entries on portfolio assessment topics found in the literature, only seven articles either reported technical data or employed accepted research methods. Similarly, only a small number of the books and articles on portfolios specifically address the use of portfolios in teacher education (Wade & Yarbrough, 1996). Measurement concerns abound including questions about the reliability and validity of portfolios as evaluation tools (Linn, Baker, & Dunabar, 1991). However, as the use of portfolios becomes more prevalent, educators are learning more about how to address these issues. For example, Naizer (1997) reported that performance portfolios are a valid method of assessing desired abilities of preservice teachers in a mathematics/science methods class. He also found that the portfolios could be reliably evaluated. Assessment issues surrounding the use of portfolios are dependent on the purpose, implementation, and context of their use. Thus, one cannot look for a blueprint or formula to lead to successful implementation, but instead processes must be designed to address these issues for each specific context. Aside from these measurement concerns, reports on the use of portfolios across a variety of contexts and programs have provided consistent claims that this assessment process has clear benefits for preservice teachers in the areas of reflection and metacognition. This assessment strategy has potential to help us transform our teacher education programs.

PORTFOLIOS IN PRACTICE

The College of Education at Wayne State University (WSU) is situated in an urban environment and enrolls approximately 3,500 students, a third of whom are in initial teacher preparation programs. Students in the teacher preparation program have from two to four semesters of pre-student teaching field experiences and student teaching experiences in pre-K–12 schools. The extensive portfolio program described here began with a pilot group of 75 students in 1991 and now has over 1000 certification students involved with portfolio development. The portfolio process was implemented for a number of reasons. Primarily it was a tool for providing students with a summative assessment through which they could reflect on their entire teacher preparation experience. It was also developed to understand how well students integrated important teaching competencies, to encourage reflection as students enter the professional world of teaching, and to provide information that could be used to improve the teacher preparation program. The portfolio process was implemented for a number of reasons. Primarily it was a tool for providing students with a summative assessment through which they could reflect on their entire teacher preparation experience. It was also developed to understand how well students integrated important teaching competencies, to encourage reflection as students enter the professional world of teaching, and to provide information that could be used to improve the teacher-preparation program. In the fall of 1992 the portfolio was made a requirement for all students in certification programs and student were required to develop and present a portfolio. By the 1993 winter semester, all certification students developed portfolios. As the program evolved, so has the quality and depth of the student portfolios.

At WSU, students are introduced to portfolios at an orientation session prior to their first pre-student teaching field experience. They are asked to begin keeping an ongoing collection of artifacts (the folio) that documents their knowledge, skills, and experiences in working with children. The students are told that they will create, from these materials, a portfolio, which they will present at the conclusion of student teaching. The presentation is often celebratory in nature as students demonstrate what they know and can do as well as how they have grown as an educator. The presentation serves as a culminating experience as they complete their certification programs and enter the teaching profession. Thus, the process is one that occurs over a period of two or three years, beginning with their first field experience and concluding after they have completed student teaching.

When the portfolio requirement was implemented, most, but not all components of the assessment system were in place. This issue has also been detailed by others (Beed & Heller, 1997) who have also recommended going forward and implementing changes as the program is analyzed and developed. Administrators of the program noted that once the portfolio became a requirement for program completion, faculty and students alike began displaying a more serious level of involvement and commitment to the process. This serious level of commitment exists even though the portfolios are not a "high stakes" evaluation measure. That is, students are required to develop and present portfolios in order to complete program requirements, but the portfolio requirement alone cannot prevent a student from completing the certification program.

BEGIN WITH THE END IN MIND

Whether portfolios are being used as an assessment tool for an individual course or for an entire program, one of the most critical aspects of the portfolio process is defining exactly what outcomes or goals the portfolio will address. When portfolios are used as assessments for specific courses, the course goals and objectives will serve as a framework for portfolio development. For example, the National Science Education Standards (NRC, 1996) provide an important source of information about the preparation our science education students need in order to be effective science teachers. When an entire teacher preparation program is considered, it is important to work from an agreed-upon philosophical base about what a beginning professional teacher should know and be able to do.

Prior to implementation of the portfolio program at WSU, the faculty had developed a list of ten teaching competencies (Snyder, Elliott, Bhavnagri, & Boyer, 1993) that reflected beliefs about effective teaching and best practice. Although these competencies have been revised over the last seven years (with the addition of two competencies and revisions to the others) and continue to be revised, the underlying beliefs of professional practice have essentially remained the same. The current competencies have been reviewed with respect to National Board Certification Standards (Barringer, 1993), the Interstate New Teacher Assessment and Support Consortium (INTASC, 1994) model standards, Entry Level Standards for Michigan Teachers (MSBOT, 1993), and standards of professional educational organizations representing various content areas. A close look at these competencies reveals keen similarities with other teacher preparation programs across the nation. In 1997, students seeking certification had to provide evidence of competency in the following twelve areas:

1. Knows the subject he/she teaches and how to teach them.
2. Organizes and implements effective instructional programs.
3. Demonstrates appropriate classroom management techniques to ensure a safe and orderly environment that is conducive to learning.
4. Stimulates creative and critical thinking in a variety of ways including the use of technology.
5. Demonstrates knowledge of human growth and development in planning for students, including those with disabilities, developmental delays, and special abilities.
6. Demonstrates a commitment to students and their learning.
7. Uses listening, speaking, reading, writing, and technological skills effectively.
8. Behaves in an ethical, reflective and professional manner.
9. Utilizes multicultural perspectives to enhance students' awareness and appreciation of diverse populations.
10. Selects appropriately from a variety of assessment strategies to evaluate student learning and to inform instruction.
11. Utilizes resources of the school district and community.
12. Communicates with parents/guardians/families.

PRESCRIPTIVE AND OPEN-ENDED COMPONENTS

The portfolio is described to Wayne State education students as a collection of materials that demonstrates the student's competence as a beginning professional. The portfolio and the presentation of the portfolio are requirements of student teaching. The students are told that it should "reflect the many dimensions of you as an urban educator including knowledge of subject matter, instructional strategies, classroom organization, management of student behavior, and evaluation of learning". They are also told that the portfolio should focus on the twelve student teaching competencies. Some components of the portfolio are prescribed. For example, students are told that it should include a resume, a statement of teaching beliefs, examples of lessons taught, and samples of their students' work. However, the structure, format, and content are left open ended so that the student can decide upon the best way to demonstrate their skills, knowledge, and abilities. The field instructor and the student's peers informally review the developing portfolio at the end of each field experience.

Each term the Office of Student Teaching sponsors an Open House for students. Graduates with exemplary portfolios are invited to display them at the Open House and answer students' questions. The Open House has become a very popular experience for students in the program. Portfolios that are modeled at the Open House are carefully selected to provide students with many different kinds of examples. They represent a wide array of organizational schemes, certification levels, and content specialty areas. There are usually examples of portfolios that are organized by competencies, chronological teaching experiences, themes, and other variations that provide a holistic view of the student's knowledge, beliefs, and skills. Although there is no single format used by all students, the portfolios assembled are similar in that they are usually reasonably small collections of materials that are carefully selected and assembled to provide evidence of competency across the twelve areas outlined above.

PRESENTATION IS INTEGRAL TO THE PROCESS

When students have completed student teaching, they present their portfolio to a two or three member panel of educators. This is a very exciting time for the students as they have completed the coursework associated with program requirements and are about to enter the professional world of teaching. Students are anxious to share their experiences with the reviewers. They prepare for the presentation as they would for a job interview, but in this case the presentation is aligned to the teaching competencies. The reviewer teams are comprised of full- and part-time faculty, teachers and administrators from nearby school districts, and faculty members from other departments across the university. Each team has at least one faculty member and one school district reviewer. Many reviewers have worked with the portfolio review program for more than one year and are familiar with the student teaching competencies and the portfolio process. All reviewers receive information on the student teaching competencies, the purpose of the portfolios, and an overview of the review process prior to presentation day. A brief orientation is given prior to the portfolio presentations.

Each presentation is approximately 30 minutes long, during which the student highlights different aspects of the portfolio. A brief explanation and suggestion sheet is given to students prior to their presentation so that they know how to prepare for the presentation and what to expect. The presentation process is partially prescriptive in that students are told to provide some basic information such as information about their certification areas, field placements, and teaching philosophy. However, most of the presentation is open-ended in that the students decide which components of the portfolio and their experiences they would like to highlight for the reviewers. Discussion, questions, and reactions from panel members follow the presentation.

EVALUATING THE PORTFOLIOS

Portfolio data was derived from two sources that are described below. The data include a rating form with items targeting components of each of the twelve competencies completed by both students and reviewers and an open-ended questionnaire completed by students at the end of the portfolio presentation. This data helped faculty in understanding our students and their experiences within the program.

Immediately following their portfolio presentations, students proceed to a post-presentation room where they complete a self-assessment rating form. During this time reviewers also individually rate each student's portfolio. The rating form used by the students and reviewers are the same. The rating form consists of 27 individual items representing each of the components of the twelve competencies listed in the Appendix. Each item was rated using a seven-point Likert scale ranging from "strongly agree" to "strongly disagree" to indicate the strength of agreement. Additionally, students are asked to complete an open-ended questionnaire. The questions were developed to ascertain the impact of the portfolio process on student understanding and reflection and to provide indicators on how the portfolio process might be improved and strengthened. The questionnaire provided an additional data source through which students could express other ideas, reflections, and suggestions that could help to inform the process.

The reviewers are also asked to recommend excellent portfolio examples for consideration in planning the aforementioned portfolio *Open House*. The portfolios are not graded. If a portfolio is judged "weak" or "marginal", the student teaching grade is deferred and the student is asked to re-do the portfolio. With well over 200 students presenting portfolios each semester, these cases are rare, averaging three to four students per semester.

Student teaching portfolios are not the only means by which the effectiveness of the program is assessed. Students are required to participate in a statewide testing program, the Michigan Test for Teacher Certification, which includes tests in the major and minor content areas. Other mechanisms are also in place to provide program assessment information. These include regular written surveys that are sent to all graduates, external and internal program reviews, and regular student assessments of the courses they complete. Additionally, faculty members are now required to submit teaching portfolios for consideration in merit-pay increases, promotion, and tenure awards.

The portfolio review process has gone through several iterations, with a steady refinement towards more systematic analysis. As noted above, the portfolio assessment has not taken on "high stakes" relevance because it is one of a number of summative measures that are used to assess competence. Furthermore, when portfolios do not effectively demonstrate comprehensive competence, students are given opportunities to work on the areas of weakness. It is believed that this has allowed the process to remain less rigid and less prescriptive. At the same time, it allows the study and refinement of the evaluative components of the process to aim for greater validity and reliability.

LESSONS LEARNED

Students who present portfolios had successfully completed program requirements. Thus, it was expected that they would successfully demonstrate the knowledge and skills targeted by the teaching competencies. We expected that these students would be rated favorably because they had been successful with all other program requirements. At the same time, we acknowledged that not all students were at the same level of proficiency across each of the competencies and that individual students would have personal strengths and weaknesses across these competency areas.

The sample was comprised of 232 students with 140 at the elementary certification level, 64 at the secondary certification level, and 26 who would be certified K–12 (2 students had missing information). Thirty-four students were science majors, with 25 elementary and 11 secondary pre-service teachers.

The mean rating for all of the reviewers (n = 56) across all of the items was 6.31 (SD = 0.58) where a rating of 7 indicated that the reviewer "strongly agreed" that the prospective teacher's portfolio presentation demonstrated attainment of all 27 components. Similarly, students believed that their portfolios adequately demonstrated their attainment of the competencies. The mean rating for all of the students across all of the items was 6.47 (SD = 0.58). The higher ratings by the students confirmed previous results that students tend to rate their portfolios more favorably than do reviewers (Stein, Elliott, & Snyder, 1998). The elementary science majors were rated highly across all items. While it helped to know that elementary science majors were able to demonstrate their skills and competencies through their portfolios, questions about the nature of their science and teaching learning experiences emerged. The review process has the potential to provide opportunities to glean further insight into the nature of the science experiences that our pre-service teachers enacted for their students.

The portfolio results also provided insights into specific program areas and potential areas to target for program improvement. For example, we found that the secondary-science majors scored significantly lower on items 11a (*Utilizing School District Resources*), and 11b (*Utilizing Community Resources*). Bearing in mind the small sample size, we interpret these results with great caution. At the same time, science education professors have now taken a close look at how and when Competency 11 (*Utilizing School/Community Resources*) is integrated into the science education program.

Questions among faculty and reviewers that arose as a result of the review process have also fostered change. The science education faculty members have had ongoing discussions on "best practice in science teaching and learning", what it encompasses, and how it can best be evidenced within a portfolio. Although there was agreement about general ideas such as fostering inquiry and actively engaging and involving students in meaningful science, our discussions also probed into more specific areas such as technology and integration. These discussions made us reflect about how these ideas were evidenced in our courses and program. Our discussions also focused on specific examples of how beginning professional teachers might demonstrate these ideas within their portfolios. With these ideas in mind, the science education faculty constructed a list of questions for reviewers that targeted best practices in science teaching. These included, What was the nature of the science-teaching example included in the portfolio? Was a hands-on/inquiry approach used? Were collaborative strategies used? Did unique science teaching experiences appear in the portfolio (i.e., field trips, computer integration, and guest speaker)?

In future iterations of the portfolio review, reviewers will be providing specific information that will detail the nature of science teaching and learning that is demonstrated in the student teaching portfolios. Reviewers will look for evidence of inquiry-based science experiences, the active engagement of students in science, the infusion of technology into science experiences, and "special" science experiences (e.g., field trips, guest speakers). As this data is collected and analyzed, it will serve to inform the program and help to focus program improvement in specific areas.

Student responses to the questionnaire have also provided valuable information about the portfolio process, its value to students, and the relationship of this assessment tool to program goals. As the student comments found below indicate, a key aspect of the portfolio process was the reflection that took place over time as student developed their portfolios (Reistetter & Fager, 1995). As students created their portfolios with the twelve teaching competencies in mind, they thought about the meaning of the competencies with respect to the meaning of the artifacts they had selected. They thought about the quality and significance of their experiences and presented the most relevant components in an organized manner in order to demonstrate professional competence.

> In constructing the portfolio I was forced to reflect on the competencies and examine how they could be applied in the classroom as well as in the portfolio. [Case 68]
>
> The process made me reflect on each competency seriously. It allowed me to think and re-think each event that I displayed in my portfolio. [Case 213]
>
> The portfolio process made the biggest impact on my understanding of child development, the various teaching strategies, and numerous classroom management plans. The process made me step back and look at how I will be as an educator and my positives and negatives. [Case 70]

From the perspective of a science teacher educator, one of the most evident outcomes of the portfolio process is that it has allowed students to view assessment in a new way. Students become aware of the role of self-assessment and reflection to learning. The process has also provided important information to help faculty in the process of program improvement. Through this process, we continually ask ourselves what we value and believe and how this is exemplified in our programs.

CONCLUSION

A great deal of criticism has been leveled at the notion of "teaching to the test". A new paradigm for assessment challenges this criticism. If the test is an authentic representation of valued goals and objectives, if it accurately reflects and assesses outcomes such as thinking, problem-solving, and communication, then "teaching to the test" would mirror the teaching of important goals and objectives. In essence, assessment and instruction would be integrated. This is an immense challenge for science education students who may not have experienced alternative forms of assessment as an integral part of their instruction. The model presented is one in which assessment measures are more authentic, relevant, and integrated with instruction than traditional methods. The model presents a way for implementing alternative assessments within a teacher education program. When science education students are assessed with performance tasks, they are demonstrating scientific knowledge and skills that they will be responsible for teaching to their own students. The infusion of portfolios as requirements for individual courses or as part of the teacher certification program gives students an opportunity to show what they know and can do within the context of their individual experiences. Through modeling best assessment practice, we teach prospective teachers to use a variety of assessment strategies that accurately reflect student learning. Changing assessment will not happen overnight. However, gradual infusion of a variety of assessment strategies into teacher preparation programs will better reflect the variety of instructional techniques that are currently used. Through these efforts, our students will be better prepared to help tomorrow's students.

REFERENCES

Angelo, T. A. (1995, November). Reassessing (and defining) assessment. *The AAHE Bulletin, 48*(2), 7-9.

Barringer, M-D. (1993). How the National Board builds professionalism. *Educational Leadership, 50*(6), 18-22.

Beed, P. L., & Heller, M. O. (1997). Portfolios: A source of information about program effectiveness. *The Professional Educator, 20*, 49-60.

Berenson, S. B., & Carter, G. S. (1995). Changing assessment practices in science and mathematics. *School Science and Mathematics, 95*, 182-86.

Collins, A. (1992). Portfolios for science education: Issues in purpose, structure, and authenticity. *Science Education, 76*, 451-463.

Council of Chief State School Officers (1997). *SCASS Science Project Consensus Guidelines for Science Assessment.* (ERIC Reproduction Service # ED 409329.)

Doran, R., Chan, F., & Tamir, P. (1998). *Science educator's guide to assessment.* Arlington, VA: National Science Teachers Association.

Duschl, R. A., & Gitomer, D. H. (1991). Epistemological perspectives on conceptual change: Implications for educational practice. *Journal of Research in Science Teaching, 28*, 839-858.

Freidus, H. (1998). *The portfolio process: Teachers and teacher educators learning together.* Paper presented at the annual meeting of the American Educational Research Association, San Diego, CA.

Green, J. E., & Smyser, S. O. (1995). Changing conceptions about teaching: The use of portfolios with pre-service teachers. *Teacher Education Quarterly, 22*(2), 43-53.

Herman, J. L., & Winters, L. (1994). Portfolio research: A slim collection. *Educational Leadership, 52*, 48-55.

Interstate New Teacher Assessment and Support Consortium (INTASC). (1994). *Model standards for beginning teacher licensing and development: A resource for state dialogue.* Draft for comment distributed by the Council of Chief State School Officers, Washington, DC.

Linn, R. L., Baker, E. L., & Dunabar, S. B. (1991). Complex, performance-based assessment: Expectations and validation criteria. *Educational Researcher, 20*(8), 14-21.

Lomask, M., Seroussi, M., & Budzinski, F. (1997, March). *The validity of portfolio-based assessment of science teachers.* Paper presented at the Annual Meeting of the National Association for Research in Science Teaching, Chicago, IL.

Lyons, N. (1998). Portfolio possibilities: Validating a new teacher professionalism. In Lyons (Ed.), *With portfolio in hand* (pp. 11-22). New York: Teachers College Press.

McLarty, J., Furtwengler, C., & Malo, G. (1985). *Using multiple data sources in teacher evaluation.* Paper presented at the annual meeting of the National Council on Measurement in Education, Chicago, IL.

Michigan State Board of Education Office of Teacher/Administrator Preparation and Certification. (1993, August). *Entry-level standards for Michigan Teachers.* Lansing, MI: Authors.

Naizer, G. L. (1997). Validity and reliability issues of performance-portfolio assessment. *Action in Teacher Education, 18*(4), 1-9.

National Board for Professional Teaching Standards (1993). *Adolescent and young adulthood: Science standards for National Board Certification* (a draft). NBTS: Author.

National Research Council. (1996). *National science education standards.* Washington, D.C.: National Academy Press.

Paris, S. G., & Ayers, L. R. (1994). *Becoming reflective students and teachers with portfolios and authentic assessment.* Washington, DC: American Psychological Association.

Reistetter, M. R., & Fager, J. J. (1995, April). *Assessing the effectiveness of the beginning teacher from a constructivist perspective*. Paper presented at the annual meeting of the American Education Research Association, San Francisco, CA.

Shulman, L. S. (1987). Assessment for teaching: An initiative for the profession. *Phi Delta Kappan, 69*, 38-44.

Shepardson, D. P., & Adams, P. E. (1996, March). *Perspectives on assessment in science: Voices from the field*. Paper presented at the annual meeting of the National Association for Research in Science Teaching, St. Louis, MO.

Snyder, J., Elliott, S., Bhavnagri, N., & Boyer, J. (1994). Beyond assessment: University/school collaboration in portfolio review and the challenge to program improvement. *Action in Teacher Education, 15*(4), 55-60.

Stein, M., Elliott, S., & Snyder, J. (1998, April). *Using teaching portfolios to assess teaching competencies across program areas*. Paper presented at the annual meeting of the American Educational Research Association, San Diego, CA.

Varvus, L. G., & Collins, A. (1991). Portfolio documentation and assessment center exercises: A marriage made for teacher assessment. *Teacher Education Quarterly, 18*(3), 13-29.

Wade, R. C., & Yarbrough, D. B. (1996). Portfolios: A tool for reflective thinking in teacher education? *Teaching & Teacher Education, 12*, 63-79.

Wolf, K. P. (1991). The school teacher's portfolio: Issues in design, implementation and evaluation. *Phi Delta Kappan, 73*, 129-136.

APPENDIX

Items for the portfolio assessment. The evaluators assess the degree to which the portfolio presentation demonstrates that the prospective teacher:

1a. *Knows the subject area content* he/she teaches.
1b. *Knows how to teach* those subjects effectively.
2a. *Organizes* effective instructional programs.
2b. *Implements* effective instructional programs.
3a. Demonstrates appropriate *classroom management* techniques.
3b. Ensures a *safe and orderly classroom environment*.
4a. Uses a variety of ways to *stimulate student creativity*.
4b. Uses a variety of ways to *stimulate critical thinking*.
4c. Uses *technology* effectively in the classroom.
5a. Demonstrates knowledge of *human growth and development*.
5b. Demonstrates knowledge in planning for students with *disabilities or developmental delays*.
5c. Demonstrates knowledge in planning for *students with special abilities*.
6a. Has a *commitment* to students.
7a. Uses *communication skills* effectively.
7b. Uses effective *writing skills*.
8a. Demonstrates *ethical* behavior.
8b. Demonstrates *reflective* practice.
8c. Demonstrates *professional* conduct.
9a. Understands *multicultural* perspectives.
9b. Integrates *multicultural* perspectives into practice.
9c. Demonstrates an *appreciation of diverse* populations.
10a. *Selects appropriate assessments* to evaluate student learning.
10b. Uses a *variety of assessment* strategies.
10c. *Uses assessment information* to inform instruction.
11a. Utilizes *school district resources*.
11b. Utilizes *community resources*.
12. *Communicates* with *parents*/guardians/families.

NEW TECHNOLOGIES AND SCIENCE TEACHER PREPARATION

Derrick R. Lavoie
Virtual Institute for Teaching and Learning Science

> As a learning tool, computers make kids adventurers and avid learners, taking them beyond the traditional walls of the schoolhouse... Teachers must be properly trained to integrate technology into the curriculum if the costly machines are to be more than fancy typewriters. (Wulf, 1997, p. 66)

INTRODUCTION

The professional development of science teachers has been given high priority in America (Advisory Committee to the NSF Directorate for Education and Human Resources, 1996; Fetters & Vellom, this volume; Thompson & Hargrave, this volume). Science education reform movements and recommendations have provided guidelines for what curriculum, content, and skills our preservice and inservice teachers should be able to impart to their students (AAAS, 1993; Aldridge, 1992; NSB, 1996). Success within the technology driven global marketplace of the 21st century will directly correlate to scientific and technological literacy (NRC, 1996, 1997; President's Committee of Advisors on Science and Technology, 1997).

> While a number of different approaches have been suggested for the improvement of K-12 education in the United States, one common element of many such plans has been the more extensive and more effective utilization of computer, networking, and other technologies in support of a broad program of systemic and curricular reform. (President's Committee of Advisors on Science and Technology, 1997, p. 11)

Using computer technology, students can share data and ideas across oceans, mathematically model a flowing stream, graph the respiration of meal worms, interact with the space telescope to discover an asteroid. They can learn to integrate video, audio, WWW sites, and high-end graphics to do laboratory reports, presentations, and research papers. The exponential growth of electronic journals in all fields, the 1000 new web sites being added every hour, and the time most of us now spend responding to e-mail provides a hint of the new millennium of global information exchange. It seems crucial that science educators concern themselves with how their students are able to utilize such technologies for teaching and learning.

A recent report found that while three out of every four U.S. public school classrooms have at least one computer designated for instruction the majority of teachers lack the necessary skills and experience to effectively utilize computers in their classrooms. Science teacher educators must therefore develop and deliver the essential technology experiences that will enable future and present teachers to utilize

D.R. Lavoie and W.-M. Roth (eds.), Models of Science Teacher Preparation, 163–176.

technology effectively at all grade levels. To meet this goal science teacher educators must address fundamental questions such as:

- What role should technology play in science teacher preparation and science teaching in general?
- Does the use of the technology improve the effectiveness of science teaching/learning? Where? When? How? Why?
- What technology experiences and skills should be acquired by preservice and inservice science teachers?
- What strategies are most effective for integrating new technologies within science teacher preparation programs?

These questions will have multiple responses depending on research, personal experience, and one's educational philosophy. The generalist perspective views technology as anything that makes the job easier, such as a pencil to assist students' writing. The social perspective considers technology to be a link between science and society. Technology education includes craft-base, vocational, high-tech, applied-science, technology-concepts, science-technology-society, and design approaches (Raizen, 1997).

In this chapter, I take a constructivist perspective on science teaching and view computer technology as a generative learning tool for enhancing student understanding and application of science concepts and processes. I begin by arguing that the application of computers in science teaching can be improved when it is based on theoretical referents involving reflection, constructivism, and information processing. I then review national standards that recommend how, when, and where to use technology for science instruction that is grounded in inquiry teaching strategies and constructivist philosophy. In the third section, I discuss examples of using the worldwide web and computer-based laboratories for science teaching and learning. In the final section, I consider practical questions for research and teaching with new technologies.

THEORETICAL REFERENTS FOR THE APPLICATION OF COMPUTER TECHNOLOGY

Reflection has become a powerful tool to improve teaching, problem solving, and learning (Canning, 1991; Russell, 1993).[1] Reflection typically involves questioning, analyzing, hypothesizing, and evaluating behaviors which result in practical modifications to one's teaching strategies (i.e., problem-solving processes). Constructivist theory asserts that learning involves the person as an active participant in the cognitive process of constructing and connecting ideas (Duffy & Cunningham, 1997). Information-processing theory considers learning as a process of constructing multidimensional cognitive networks that establish connections between new knowledge and previously existing knowledge. Relative to science instruction, the connections between content and process knowledge are most important.

> Improvement in problem-solving ability (i.e., learning) is viewed as a process by which connections are established and networked between procedural and declarative knowledge in increasingly complex ways—ways that lead to more effective solutions. (Lavoie, 1995, p. 14)

Computer technology offers new ways to create learning environments that support reflection and construction and promote the use of higher-level cognitive information-processing skills, scientific inquiry, and the construction/representation of conceptualized and interconnected knowledge bases. Computers embody considerable power to facilitate knowledge representation and learning:

> Computers also have the capability of creating dynamic, symbolic representations of non-concrete, formal constructs that are frequently missing in the mental models of novices. More importantly, they are able to proceduralize the relationships between these objects. Learners work out differences between their incomplete, inaccurate mental models and the formal principles represented in the system. (Kozma, 1991, p. 199)

Technology facilitates learning most when learners actively use it (Kozma, 1991). The greater the level of interactivity between the user and the technology the more involved learners can become in choosing multiple cognitive pathways resulting in more meaningful learning. Interactivity is viewed as important variable affecting learning outcomes such as conceptual understanding and the process of learning (learning how to learn). Interactivity can take several forms in a technology enriched classroom. Local networks may connect computers to a server, to each other, or to the teacher. Web-based interactive learning environments connect students, teachers, and information in various ways. Interactive discussions may occur in virtual classrooms in real time. Microworlds and highly interactive CDs allow students to simulate scientific processes and concepts, search massive databases, and in effect, create or construct their own learning pathways. It has been suggested that the computer's essence is its universality, its power to simulate, because it can take a thousand forms, and can serve a thousand functions, it can appeal to a thousand tastes (Papert, 1984). Concept mapping computer tools can be used to build visual relationships between concepts (Fisher, 1990). Hypermedia-based technologies facilitate knowledge construction by enabling multiple connections through electronically linked node networks. Thus,

> The use of video and audio nodes enables the presentation of environments rich in information in a real-world context, and enhances flexibility in recalling information ... the electronic links between various modes of information representation help the formation of meaningful links between existing knowledge and new knowledge, and facilitates comprehension by encouraging students to explore new information in multiple perspectives. (Kumar & Sherwood, 1997, p. 250)

Considering the variety of ways computer technology can be applied for learning and teaching and the rapid evolution of computer technology we can expect dramatic transmutations in how teachers teach and how they are prepared to teach science. This will shift education toward the constructivist paradigm, as the traditional role of the teacher as the "all and knowing" must change toward the teacher as the facilitator or "coach" of knowledge construction.

> The new technology will transform the role of the teacher as thoroughly as did the introduction of printed textbooks. More than in the past, teachers must become advisors to student inquirers, helping them to frame questions for productive investigation, directing them toward information and interpretive sources, helping them to judge the quality of the information they obtain, and coaching them in ways to present their findings effectively to others. This will require teachers to become even better prepared in the

> content of the subjects they teach, and the means by which the content can be taught and learned. (Task Force on Technology and Teacher Education, 1997, p. 4)

Collins (1991) identifies several constructivist pedagogical shifts resulting from the use of new technology such as increased tendencies for: small-group work, coaching, cooperative learning, student engagement, integrated visual and verbal learning, individual learning goals, and assessment strategies based on products, progress, and effort. The expansions of new technology in schools will put greater emphasis in science education on constructivist learning where "learning to learn" and "dealing with change" will become the most important goals (Hurd, 1991). Of course, science educators will have to prepare future teachers to develop pedagogical understandings compatible with technology use. The technology-concerned recommendations of the national standards provide further insight into the goals and directions for incorporating new technology in science education.

NATIONAL STANDARDS AND RECOMMENDATIONS FOR APPLICATION OF TECHNOLOGY IN SCIENCE TEACHING

To effectively use new technologies science teachers must have guidance, appropriate experiences, and support. The recently released *Standards for Science Teacher Preparation* (NSTA, 1998) provide general recommendations for what teachers should be able to do with technology but provide limited guidance in terms of specific strategies or methods by which to accomplish the recommendations.

- Science teachers should use technology extensively in their preservice curriculum and be provided opportunities to know and understand applications of science in community and workplace environments.
- Teachers must be able to use technology effectively for the level they are preparing to teach. All science teachers should be able to use computers to provide simulations, games, and information.
- At the secondary level, teachers should be able to use computers to collect and process data, and to install and use peripheral sensing devices for investigations.
- Technology goes beyond computers, however. Teachers should be able to smoothly incorporate multimedia and audio-visual technology into their teaching, and should involve their students in doing so.

The *National Science Education Standards* consider technology to be a "tool" to assist with developing scientific understanding through inquiry (NRC, 1996). Thus, "Effective science teaching depends on the availability and organization of materials, equipment, media, and technology.... Teachers provide the opportunity for students to use contemporary technology as they develop their scientific understanding" (p. 44). Both the *Standards for Science Teacher Education* and the *National Science Education Standards* emphasize the relationship between science, technology, and society. A common thread of agreement running through all the standards and recommendations is that pre-service and inservice science-teacher preparation programs incorporate technology in a variety of ways and that doing so should enhance teaching and learning in the science classroom. Research demonstrates that use of technology in science teaching improves motivation (Kumar, 1997), visualization, concept formation (Fisher, 1990), scientific processing (Mokros, 1987;

Stuessy, 1989), scientific problem solving (O'Connor, 1994), and scientific attitudes (Harwood, 1997).

Some possible technology applications include (President's Committee of Advisors on Science and Technology, 1997, p. 36):

- An information retrieval or database search engine capable of extracting information from a single system or from sites distributed across the global Internet.
- An environment for the simulation of any of a wide range of devices and machines, physical systems, work environments, human and animal populations, industrial processes, or other natural or artificial systems.
- A facility for the collection, examination, and analysis of statistical data (which might be used in connection with any of a wide range of experimental or survey applications).
- An environment for domain-specific problem solving.
- An environment for the facilitation of group collaboration.
- A flexible laboratory instrument supporting the collection of scientific data from various sensors and the flexible manipulation of this data under student control.
- A medium for communication with teachers, parents, community members, experts, and other students, both locally and over great distances, and for the organization and coordination of group projects.

In sum, the national guidelines provide general recommendations for the use of technology in the science curriculum but without the specificity of how such applications can take place in K-12 classrooms or within science teacher preparation programs.

RESEARCH-BASED STRATEGIES USING COMPUTER TECHNOLOGY IN THE PREPARATION OF SCIENCE TEACHERS

In this section, I discuss several research-supported efforts that used new technologies as part of the science teacher preparation program at Black Hills State University (BHSU). Linking new technology with effective constructive-based strategies not only utilizes the technology most effectively but also may create a synergistic improvement in scientific problem-solving skills and conceptual understanding. Specific applications employing on-line collaboration via the WWW and computer-based laboratories for inquiry science are described.

Electronic Discussion Forums

Science teacher educators should train teachers to utilize the Internet for effective discussion and collaboration concerning a variety of topics in science teaching and learning (Robinson, 1994; Shepardson, 1995). Written ideas, quantitative data, and interactive graphics can be created, posted, debated, and evaluated to establish a dynamic and flexible knowledge network. Berger, Lu, Belzer, and Voss (1994) recognize the potential of on-line discussions for the active construction of knowledge. Thus, "Critical thought and doubt are enhanced as students realize that justification for their ideas is crucial in determining how uncertain their position is and whether their "knowledge" should be tested further or perhaps changed" (p. 481). Based on quantitative and qualitative analysis of questionnaire, interview, and email logs, Lavoie (1997) made the following comment concerning the effects of teaching university science content/education courses to high-school science teachers via email.

> The virtual electronic learning environment was shown to be an effective means of providing useful teaching/learning experiences to in-service high-school science teachers on an on-going basis during the school year. Courses were most successful when collaborative and constructivist learning environment was established in which the science teacher participants exercised control over the direction of their learning. New strategies for teaching with distance education must continue to be sought as new technologies emerge and change the nature of the traditional classroom. (p. 9)

Lavoie and Foster (1996) examined the dynamics of using e-mail collaboration between students at different universities to enhance reflective thinking in science methods courses. Their students gained confidence in using e-mail as a tool for reflecting about scientific concepts. Establishing a meaningful purpose about which the participants can exchange and construct useful ideas is also important. The following recommendations should be considered regarding the use of electronic discussions (Lavoie & Foster, 1996, p. 17):

- Provide adequate time for orientation. During this time students must establish links between all members of a collaboration cohort and the instructor and develop a cohort distribution list. This will probably take at least a week as some students will not be as familiar with technology as others and will need training.
- Establish discussion guidelines. For example, this might involve setting a date by which all members of a cohort must have read the messages of the cohort and replied at least twice. This will force students to establish a regular time-line for getting on the system and interacting, which should be at least once a day.
- Establish assessment criteria for quantity and quality of interaction. This will not only motivate the students to interact but will provide the instructors with feedback that can be used to modify the student activities.
- Provide a relevant context for collaboration. Students tend to choose more concrete and familiar topics such as lesson plans, classroom management, and science teaching resources over more abstract or theoretical topics such as constructivism.
- Maintain flexibility and remember that "Murphy's law" applies to technology, perhaps more so than it should.
- Provide some class time for on-line discussions rather than trying to do it only as an after class assignment.
- Develop similar goals, objectives, assignment guidelines, and assessment criteria that integrate across participating universities. Every attempt should be made to work concurrently between universities relative to time and depth.

In sum, teacher education students can use electronic discussions to share and debate the importance and analysis of scientific data and information, experimental design, prediction and hypotheses, and apply a variety of other scientific processing and inquiry skills. In practice, discussion assignments may center on pedagogical issues in science teaching such as assessment, classroom management, and constructivist learning. Discipline-focused assignments can involve the students questioning and discussing locally relevant phenomena such as the phases of the moon, water pollution in the nearby creek, or a recent outbreak of mononucleosis.

Hardware and Software

Advances in computer technology make it easy for science teacher educators to engage students in interactive on-line discussion via the world-wide-web (WWW). This typically involves WWW server software that can turn most of the newer PC or Macintosh computers into a WWW server computer assuming the connection is via an assigned IP (internet protocol) address. The server software enables the host computer to receive and send various types of information to other computers on the WWW. The server software also facilitates the presentation of dynamic web pages created by separate software. Additional software is needed to facilitate and organize interactive on-line discussion. Software for synchronous (real-time) discussion such as IRC (Internet Relay Chat) software allows many users on different systems at different locations to converge into one "room" for real-time interactive discussion, similar to a conference call. Software for asynchronous (lag-time) discussion includes various usenet groups, LISTSERVE, listproc, mailserv, majordomo, LOTUS notes, and INTERACTION. Much of this software is available for all platforms and can be obtained as freeware or shareware from such sites as TUCOWS (http://tucows.tierranet.com/) or DOWNLOAD (http://download.com/).

Science Methods Class Example:

The following example reviews on-line discussions that recently took place as part of an integrated discussion between elementary and secondary science methods courses at BHSU. The technology components included a Macintosh G3, WEBSTAR WWW server software, and INTERACTION discussion software. INTERACTION provides a friendly user interface with icon menus to allow easy entrance into a discussion topic, posting of new discussion topics, and inputting of threaded responses to a discussion topic.

An instructional strategy, which involves a motivational focus, increases cognitive demand, and increases interactivity, was used to engage the students in on-line discussions. These discussions took place throughout the methods courses. At the beginning of each course students were given two technology orientation sessions, which covered the use of email, course web site operation and development, and the use of the discussion forum. This was done in a dedicated computer room that allowed each student to operate their own computer. An initial on-line discussion took place during one of these orientation sessions in class to reduce frustration and identify and "bugs" associated with using the on-line discussion forum. The students were to reply to a posted question asking them to describe why they wanted to teach science. To ensure a greater quality of responses the students prepared and edited their work in a word processor before copying and pasting into the discussion forum. Most computers on the market today run "Windows" with enough RAM (16 megabytes) to allow the word processing application to be open at the same time as the WWW browser (e.g., Netscape™) is open. Users simply click back and forth between the different applications to copy and paste. A second more pedagogically focused discussion question asked students to describe and validate past teaching experiences in which they had learned science concepts. The WWW interface for this question is shown in Figure 1.

A second discussion question concerned the issue between creationism and evolution (Figure 2). To increase interactivity, students were told that to receive a passing grade they must upload and initial response and make at least one reply to someone else's initial response. There were over 150 responses in this threaded discussion! A third discussion item was more content focused.

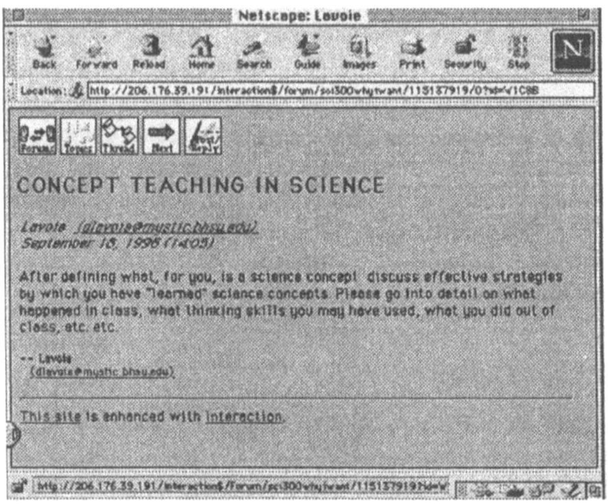

Figure 1. Posted discussion question concerning science pedagogy.

One of the exchanges went as follows.

September 24, 1998 (13:49)
Creationism is a philosophy, or to be more accurate, it's a part of a particular Cosmology. This posture is very personal, and held by a certain segment of western culture as we know it. Other than the category of "additional perspectives", in any Science classroom, concerning any area, this Philosophy should not be addressed in the public school setting. Jefferson and the balance of the founding fathers of this country, wrestled with this issue more than any other in their deliberations. The profound wording in that document states that we have two basic protections----protection of an individuals religion, and (equally important)----from religion. It is my position that these basic tenets apply in the public domain. A crucial component of that domain is the public school system. Doug (ddddd@university.edu)

September 26, 1998 (08:43)
I am not a member of any specific religious faction or group at this time and I am a firm believer in separation of church and state as outlined in the Constitution. Science consists of theories based in fact whereas Religion consists of theories based on speculative beliefs and faith. The creationism theory doesn't have a place in a state or federally supported institution. If approached, I will simply tell my students that Evolution is one theory. If they wish to learn about creationism, they should consult a member of the clergy and I believe that they should get both sides to be fully informed. Steve (steve@university.edu)

September 29, 1998 (20:08)
I think that as the racial and ethnic make-up of this country continues to change, that we have not only a desire to learn about different beliefs but also, in fact, an obligation to do so. Evolution and creationism are an old argument, in my opinion. As educators approaching a time in this country unprecedented by ethnic mixing we need to be diverse in the information that we offer. Not all cultures ascribe to either one or the other belief. When teaching this subject why not open with questions of how other cultures have answered this question. Especially in S.D. we could touch on the Lakota story of the White Buffalo Woman, or another Native American belief. Yes, the theory of evolution should be taught as a possible explanation, but why not also let your class be less Anglo centric and more broad minded? What are you afraid of? Betsy (betsy@university.edu)

October 1, 1998 (22:16)
(Steve replies to Betsy) I am afraid of a state in which somebody's private religious beliefs are forced upon my family or me. I don't think that I can agree with your argument that we should teach religious creationism along with scientific evolution. The greatest problem would be fairness...for instance you mention the possibility of teaching Lakota creationism to a South Dakota class. What if the class consists of eight Anglo-Saxons, four African-Americans, one Pacific Islander, and three Asian Americans. Should they have thrust upon them by an intrusive state, the religious beliefs of a culture in which they may have limited or no interest? Each may (or may not) desire to see their particular religious belief covered. I for one would insist that if somebody else's cultural beliefs were covered then mine must be too. While we are covering each of our beliefs, or the Lakota beliefs, we are ignoring hundreds of other beliefs, each of which has its adherents. Its not that I don't believe religious creationism has its place... its just that I don't think we should use the power of the state to impose it on any given person in this manner. What would your reaction be if the state imposed a mandatory course on any particular religious beliefs? Steve (steve@university.edu)

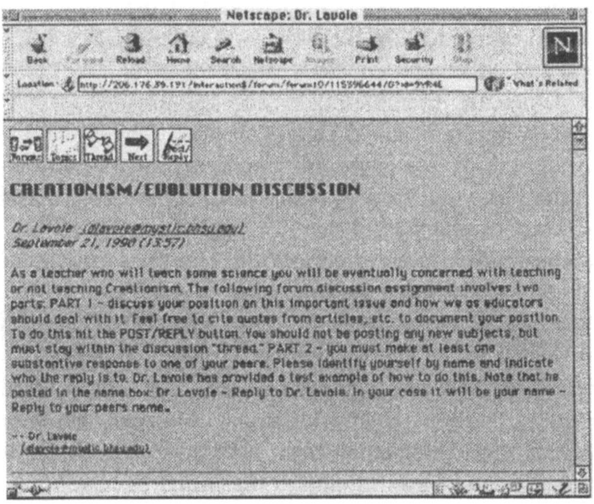

Figure 2. Posted discussion assignment to initiate the creationism/evolution discussion.

Post-course questionnaire data indicate that the majority of students were positive about using the web site and taking part in the on-line discussions. Students felt that the discussions allowed them to significantly broaden their understanding of a particular concept or issue. Many were appreciative of the flexibility that the Internet discussion assignments gave to their schedules. On the negative side, students felt that it was sometimes difficult to gain access to the web site and it was difficult for some to get access to networked computer. Some students who were less experienced with technology felt they needed exorbitant amounts of time to contribute to the discussion assignments to simply learn "how to do it". Some students did not wish to enter personal phone numbers in the web site database. As security does become more of an issue, WEBSTAR has the capability to moderate the discussions and other parts of the web site using password access.

In summary, electronic discussions are most effective when they facilitate the construction of knowledge. The nature of the discussion question or topic must engage the students to read others responses and to make their own responses. Through electronic interaction, teachers and students can become part of learning community in which they can share ideas and data, get advice, tap resources, collaborate on projects and experiments, and extend the envelope of effective science teaching and learning.

Computer-Based Laboratory

Computer-based laboratory (CBL) is a general term used to describe a system composed of scientific interface probes that are coupled through an analog-to-digital converter to a CBL unit that is in turn connected to a computing device. Probes are connected to the CBL unit through multiple input or output connections that are called channels. Sensor probes can collect motion, temperature, light, sound, pH, force, heart rate, and even EKG data. Data can be collected at a rate of up to 10,000 points per second and on several channels simultaneously. The computer is used to collect, analyze statistically (parametrically and non-parametrically), graph, model, and display a variety of "real world" scientific data. The TI-83/85 (Texas Instruments [http://www.ti.com]) is a powerful "intelligent" hand-held computing device that connects to the CBL unit (e.g., Vernier [http://www.vernier.com]). As data is collected through probes attached to the CBL unit it is sent to TI-83 and stored in lists. The technology provides an important link between mathematics and science as data is displayed, manipulated, and modeled to reveal scientific processes. Experiments that were out of reach just a few years ago can now be done quickly, easily, and safely in the lab and in the field.

CBL is a relatively new and inexpensive tool that can be used effectively at many grade levels to facilitate scientific inquiry and scientific understanding. For example, students in upper elementary and middle school can use the CBL to collect, analyze, and display scientific data in real time. Research has shown that by observing graphs being formed in real time students develop and understand the relationships between variables and not just a static set of points (Nachmias & Linn, 1987). CBLs facilitate more accurate and efficient data collection, thus freeing up more classroom time for scientific processing (hypothesizing, predicting, designing experiments, reaching conclusions, synthesizing, etc.). In general, research has shown that using technology tools such as the CBL leads to more positive science attitudes, greater student ownership, increased scientific understanding, and generally promotes learning partnerships and constructivist teaching (Brassell, 1987). Of course the key factor is the ability of the teacher to use the CBL as a tool to promote concept and process learning, not the CBL *per se*. The following brief example illustrates how CBL/TI-83 can be used as part of inquiry-science activity based on a learning-cycle strategy.

Exploration: Students initially predict and explain how skin temperature and internal body temperature will change following exposure of the body to freezing temperatures and vigorous exercise. Using the CBL and two temperature probes, students explore how their own skin and internal body temperatures change, simultaneously, in response to hot and cold temperatures and draw inferences concerning the concept of homeostasis.

Term Introduction: Homeostasis is defined and some examples are given of animals that do (e.g., mammals) and do not (e.g., insects) sustain constant internal body temperatures. Students are led to realize that their body systems adjust in a variety of ways to abiotic changes in the environment to maintain a steady state inside their bodies.

Concept Application: Students discuss hypothermia and additional adaptations that warm-blooded animals use to maintain a steady internal environment (e.g., fur, burrows, blubber, wool, and feathers). Students design various types of insulation materials and develop strategies that would keep them alive if they were to be stranded in a winter snowstorm.

NEW DIRECTIONS FOR TECHNOLOGY IN SCIENCE TEACHING AND RESEARCH

New technologies can "provide opportunities for teacher education students to compare classroom episodes of effective and ineffective science teachers and analyze critical components of instructional strategies" (Kumar & Sherwood, 1997, p. 258). Educational research shows that the use of advanced computer technology leads to increased motivation, greater task attention, greatly increased visualization potential (interactive 3-D graphics), and generally increased comprehension of the subject matter (Bates, 1994; Peterson, 1995). One advantage of computer technology rests with its capacity to construct and deliver instruction involving complex content and concepts in the sciences and provide a medium by which students can construct and learn complex concepts and ideas in the sciences.

While computer technology does offer many opportunities for enhancing learning, teachers and teacher educators must be concerned with whether the use of technology in particular instances will improve or actually reduce learning, and in particularly concept or meaningful learning (Ausubel, 1986). Technology facilitates the exploration and linkage of information in non-linear ways to create flexible cognitive networks that can be used for more effective learning and problem solving. In a review of research in the uses of technology in science education Berger et al. (1994) reported positive effects for science learning when using microcomputer-based laboratories (Brassell, 1987) and interactive videodisc. However, there is a lack of research concerned with the use of computer technologies for meaningful science teaching and learning. Empirical research studies need to follow exploratory and formative studies to address the above questions to validate the success or failure of computer-based technology as a tool to facilitate the construction and integration of knowledge.

The following research questions should set the pace for future research in science education researchers considering the applications of new technologies for science teaching and learning.

- What is the qualitative and quantitative nature of knowledge construction using new technologies?
- How can the new technology be used to facilitate the generation of new understanding for under-served populations?
- How can new technology be used to help students develop conceptual understandings of science and improve the use of cognitive-processing skills.
- What is the effect of technology-based instructional strategies that promote student control over their own learning?
- What strategies are most effective for collaborative on-line interaction?
- How can the electronic medium be made more personable?
- What strategies are most effective when a teacher has one or two computers in a classroom or when every student has a computer?
- What strategies are most effective for assessing students learning and teaching using various types of technology? How can exorbitant time demands to learn new technologies be reduced for the instructors and the students?

On a more practical level, science teachers must be cautious about the use of technology in their classroom and evaluate its use against relative to cost, effectiveness, and educational goals and objectives. They have to ask themselves questions such as Can I use this technology to teach in the way that I want to teach better? Can I teach the concepts and processes I want to better without it? Will I be able to easily train my students to use it? What access will my students have to this technology after they graduate? How long will this technology be available? How much will it cost for me or for my entire class? What kind of "extras" do I need to buy? What kind of technical support do I need or can I expect?

CONCLUSION

It seems inevitable that advances in computer technology will transform how teachers teach and students learn—and therefore, how we go about preparing future science teachers. At the recent 1998 National Educational Computing Conference in San Diego, futurists spoke of miniaturized body computers that will fit in the eyes, ears, and brain. These body computers will enable students, scientists, and teachers to communicate with each other instantaneously anywhere over the globe, access unlimited databases of information, and build new worlds in virtual space. Students will no longer be tied to one teacher who must stand and deliver. Rather, students will encounter multiple forms of auditory, visual, and kinesthetic knowledge and (synthetic and biological) information that will guide them through unlimited learning pathways.

Technology can be used most effectively in science teaching as a tool to enhance the implementation of other teaching strategies such as inquiry, learning cycle, and cooperative learning. The ultimate goal of good science-teaching strategies is to improve conceptual understanding and processing skills. Berger *et al.* (1994) suggested that technology has and will continue to shift the focus of science education research toward constructivist-based research themes. Still further, I believe that such a shift will fundamentally transform the nature of science teaching and learning. This shift will move us away from didactic "sage on stage" traditional approaches to student-centered open-ended inquiry approaches.

The use of new technology adds to the already heavy burden of responsibility for science-teacher educators who must prepare science teachers for the future not the present. Training K-12 teachers how to effectively utilize technology to ensure that their students conceptualize, process, apply, and like science is, indeed, an "awesome" task. As we move into the new millennium, science educators can best utilize computer technology as a tool for the application of constructivist learning and teaching strategies. But, technology can be saving grace or a destructive nemesis and is certainly not to be perceived as the solution to our educational problems. When the use of technology is contrary to basic education goals it should not be used. Ultimately, what science teachers are able to do with technology does not rest with the technology but with each of us.

Finally, I believe that science teacher educators cannot ignore or be intimidated and controlled by technology. Rather, science teacher educators must embrace technology and adapt it to personal and national needs for teaching and learning.

NOTES

[1] For a critical view of regarding the notion reflection see the chapter by Roth.

REFERENCES

Advisory Committee to the NSF Directorate for Education and Human Resources. (1996). *Shaping the future: New expectations for undergraduate education in science, mathematics, engineering, and technology*. Washington, D.C.: National Science Foundation.

Aldridge, B. (1992). Project on scope, sequence, and coordination: A new synthesis for improving science education. *Journal of Science Education and Technology, 1*, 13-21.

American Association for the Advancement of Science. (1993). *Benchmarks for science literacy: Project 2061*. New York: Oxford University Press.

Ausubel, D. (1986). *Educational psychology*. New York: Holt, Rinehart, and Winston.

Bates, A. W. (1994, June). *Educational multimedia in a networked society*. Paper presented at the World Conference on Educational Multimedia and Hypermedia, Vancouver, B.C., CANADA.

Berger, C. F., Lu, C. R., Belzer, S. J., & Voss, B. E. (1994). Research on the uses of technology in science education. In D. L. Gabel (Ed.), *Handbook of research on science teaching and learning* (pp. 466-488). New York: Macmillan.

Brassell, H. (1987). The effect of real-time laboratory graphing on leaning graphic representations of distance and velocity. *Journal of Research in Science Teaching, 24*, 385-395.

Canning, C. (1991). What teachers say about reflection. *Educational Leadership, 48*, 18-21.

Collins, A. (1991). The role of computer technology in restructuring schools. *Phi Delta Kappan, 73*, 28-36.

Duffy, T. M., & Cunningham, D. J. (1997). Constructivism: Implications for the design and delivery of instruction. In D. H. Jonassen (Ed.), *Handbook of research for educational communications and technology* (pp. 170-198). New York: Macmillan.

Fisher, K. M. (1990). Semantic networking: The new kid on the block. *Journal of Research in Science Teaching, 27*, 1001-1018.

Harwood, W. S. (1997). Effects of integrated video media on student achievement and attitudes in high school chemistry. *Journal of Research in Science Teaching, 34,* 617-31.

Hurd, P. D. (1991). Why we must transform science education. *Educational Leadership, 49,* 33-35.

Kozma, R. B. (1991). Learning with media. *Review of Educational Research, 61,* 179-211.

Kumar, D. (1997). Computer technology, science education, and students with learning disabilities. *Journal of Science Education and Technology, 6,* 155-60.

Kumar, D., & S. Sherwood, R. (1991). Hypermedia in science and mathematics: Applications in teaching education, problem solving and student testing. *Journal of Educational Computing Research, 17,* 249-256.

Lavoie, D. R. (1995). The cognitive-processing nature of hypothetico-predictive processing. In D. R. Lavoie (Ed.), *Toward a cognitive-science perspective for scientific problem solving* (pp. 13-50). Manhattan, Kansas: Kansas State University Press.

Lavoie, D. R. (1997). Use of telecommunications to deliver university science content/education courses to high-school science teachers: an evaluation. *Electronic Journal of Science Education, 1,* Available: http://unr.edu/homepage/jcannon/ejsev1n3.html.

Lavoie, D. R., & Foster, G. (1996, January). *An inter-university internet exchange project to network pre-service science teachers.* Paper presented at the annual meeting of the Association for the Education of Teachers in Science, Seattle, Washington.

Mokros, J. R. (1987). The impact of microcomputer-based labs on children's ability to interpret graphs. *Journal of Research in Science Teaching, 24,* 369-383.

National Research Council. (1996). *National science education standards.* Washington, D.C.: National Academy Press.

National Research Council. (1997). *Preparing for the 21st Century: Technology and the Nation's future.* Washington, D.C.: National Academy of Sciences.

National Science Board. (1996). *Science and engineering indicators.* Washington, D.C.: U.S. Government Printing Office.

National Science Teachers Association. (1998). *NSTA standards for science teacher preparation.* Washington, D.C.: National Science Teachers Association. Available online: http://www.iuk.edu/faculty/sgilbert/nstastand98.htm

O'Connor, J. (1994). The effects of technology infusion on the mathematics and science curriculum. *Journal of Computing in Teacher Education, 10,* 15-18.

Papert, S. (1984). *Mindstorms: Children, computers, and powerful ideas.* New York: Basic Books.

Peterson, N. K. (1995). Implementing multimedia in the middle school curriculum: Pros, cons and lessons learned. *T.H.E. Journal,* 70-75.

President's Committee of Advisors on Science and Technology. (1997). *Report to the President on the use of technology to strengthen K-12 education in the United States.* Washington, DC: Executive Office of the President.

Raizen, S. A. (1997). Making way for technology education. *Journal of Science Education and Technology, 6,* 59-70.

Robinson, M. (1994). Using email and the Internet in science teaching. *Journal of Information Technology for Teacher Education, 3,* 229-38.

Russell, T. (1993). Learning to teach science: Constructivism, reflection, and learning from experience. In K. Tobin (Ed.), *The practice of constructivism in science education* (pp. 247-258). Hillsdale, New Jersey: Lawrence Erlbaum Associates.

Shepardson, D. P. (1995). Mathematics and science teaching and learning on the information superhighway. *Journal of Computers in Mathematics and Science Teaching, 14,* 9-26.

Stuessy, C. (1989). Advantages of micro-based labs: Electronic data acquisition, computerized graphing, or both? *Journal of Computers in Mathematics and Science Teaching, 8,* 18-21.

Task Force on Technology and Teacher Education. (1997). *Technology and the new professional teacher: Preparing for the 21st century classroom.* Washington, DC: NCATE.

Wulf, S. (1997). Teach our children well (it can be done). *Time, 150,* 62-69.

PREPARING NEW TEACHERS FOR INTEGRATED-SCIENCE CLASSROOMS

Robert Yager, Sandy Enger, & Ann Guilbert
University of Iowa

INTRODUCTION

At the heart of this chapter is the preparation of teachers to teach science in an integrated manner as developed in the Scope, Sequence, and Coordination (SS&C) project (National Science Teachers Association, 1992). This project was conceived and sponsored by the National Science Teachers Association with major support from the National Science Foundation and auxiliary support from the U.S. Department of Education, the American Petroleum Institute, and a variety of industries in six states and several hundred schools in Alaska, California, Iowa, North Carolina, Puerto Rico, and Texas. The SS&C project focused on a required integrated science sequence of middle and high schools courses. This sequence was to change the typical "layer cake" sequence (i.e., grade 10 biology, grade 11 chemistry, and grade 12 physics) and the discipline-bound junior high schools programs (i.e., life, earth, and physical science). This project aimed to include some of each science discipline each year of schooling which respected major themes and connections among the sciences, and which were situated in relevant, local, and current problems. This new integrated science effort also focused upon new goals appropriate for all students, new instructional strategies, and especially new models for assessing success. With funding that began in 1990 and ended in 1997, the Iowa project focused upon these basic "incidents" of school programs in grades 6 through 10. The activities were all school-based and involved in-service teachers. However, preparing new teachers for integrated science as defined by the participating districts was a major effort of the science teachers education program at the University of Iowa. Unfortunately, funding was curtailed in 1997 thereby abandoning one of the central goals—altering the biology, chemistry, and physics sequence of courses and providing a model for implementing a complete 6–12 integrated science program envisioned in the *National Science Education Standards* (National Research Council, 1996). In California many of the high schools began with initial changes and continue with an integrated science approach today (Scott, 2000).

In this chapter, we present the history and context of teaching science in an integrated way in Iowa schools and the associated efforts at the University of Iowa to prepare teachers for teaching science in an integrated way. We then outline some of the fundamental elements in science teacher preparation. Two case studies are used to exemplify the kind of science teaching enacted by graduates from the program.

D.R. Lavoie and W.-M. Roth (eds.), Models of Science Teacher Preparation, 177–194.

HISTORY AND CONTEXT OF INTEGRATED SCIENCE IN IOWA

In Iowa, efforts began in 1990 to involve all pre-service students in the reform and assist them to do their student teaching in districts involved with the major reforms. The twenty participating "districts" were organized in four centers, including North Iowa with headquarters in Mason City, Southern Prairie with headquarters in Chariton and Ottumwa, Central Iowa Science Network with headquarters in Creston and Johnston, and the Davenport Community School District. These four centers became actively involved in the preparation of new science teachers and thereby became professional development centers. (See also the chapters by Barufaldi/Reinhartz and Fetters/Vellom on how such collaborations can be made to work.) These schools were more typical than those in Iowa City and others nearby where the three practicum experiences occurred as a part of a methods sequence over a three-semester period.

Preparation of new teachers ready for teaching integrated-science curricula is critical to reform efforts. The national project represented one of the largest reform efforts undertaken in the United States in terms of funding, number of teachers involved, and impact on whole districts. Although the funding effort was terminated in 1997, preparation for teaching science in an integrated way remains central to the secondary pre-service sequence at the University of Iowa and other institutions with major teacher education programs.

The major outcome of the Iowa integrated-science program was the promotion of more and better student learning in six domains. The six learning domains include concepts, process, application, creativity, attitude, and worldview.

Data are available from the classrooms of new teachers as well as the inservice teachers who were centrally involved with the NSF reform project. Several dissertations establish the effectiveness of an integrated science approach and the successes of new teachers from the Iowa teachers education program (Craven, 1997; Hollenbeck, 1999; Kimble, 1999; Krajcik, 1986; Stiles, 1993; Tillotson, 1996). A basic assumption for successful teaching (and a precursor to learning) is the intellectual engagement of all students. Teaching strategies are deemed more important than a particular curricular pattern. Therefore, much time is spent in analyzing teaching in addition to the re-alignment of science content. Both, however, are important if students are to be intellectually involved to the degree needed for learning. Perrone (1994) identifies eight ways in which such student mind engagement occurs. These are:

1. Students help define the content.
2. Students have time to wonder and to find a particular direction that interests them.
3. Topics have a "strange" quality—something common seen in a new way, evoking a "lingering question".
4. Teachers permit—even encourage—different forms of expression and respect students' views.
5. Teachers are passionate about their work. The richest activities are those "invented" by teachers and their students.
6. Students create original and public products; they gain some form of "expertness".
7. Students *do* something—e.g., participate in a political action, write a letter to the editor, or work with the homeless.

8. Students sense that the results of their work are not predetermined or fully predictable.

All eight points are used as teaching and curriculum are evaluated. Once intellectual engagement is realized, learning is likely to occur. Since real learning rarely occurs as a result of the type of instruction that characterizes most K–12 and college science classrooms, new approaches are needed (Brooks & Brooks, 1993; Lochhead & Yager, 1996; Mestre & Lochhead, 1990). Typical teaching places premiums on students who can repeat, recall, and regurgitate science concepts and process skills that provide the curriculum frameworks in most settings. As late as 1990, the goals for science education have been defined in terms of two dimensions, namely concepts and processes (Mestre & Lochhead, 1990). But this two-dimensional view of science is not adequate for the current reform initiatives; these two represent only part of the essential science content outlined by the *National Standards*. More important than typical science content is the issue of how teachers teach and what students are expected to learn.

In our teacher preparation for integrated science, we assume that integrated-science teachers must be reflective and must think of their teaching as a science itself. They must question their behaviors and actions and hypothesize about how they might impact learning both negatively and positively. They must raise questions for which possible answers can provide the basis for observations, data collection, and data analysis to determine the validity of the idea (original explanation) that was proposed. This collection of evidence concerning the validity of personally constructed explanations exemplifies science, constructivist thinking, and the teaching actions required in integrated science classrooms. The specific procedures have been defined and examples provided elsewhere (Lochhead & Yager, 1996; Yager, 1995, 2000).

The Iowa project was conceived as a reform effort that illustrates the visions outlined in the *National Science Education Standards* (NRC, 1996). The *Standards* call for teachers who can (p. 6):

1. Plan an inquiry-based science program for their students.
2. Guide and facilitate learning.
3. Engage in ongoing assessment of their teaching and of student learning.
4. Design and manage learning environments that provide students with the time, space, and resources needed for learning science.
5. Develop communities of science learners that reflect the intellectual rigor of scientific inquiry and the attitudes and social values conducive to science learning.
6. Actively participate in the ongoing planning and development of the school science program.

But the *Standards* also emphasize that teaching must be coordinated and result in a program or curriculum. Basic ingredients for achieving an exemplary program include (p. 7, 8):

1. Assuring that the K–12 science program is consistent with other National Standards and that the NSES visions are articulated within and across grade levels to meet a clearly stated set of goals.
2. Developing a curriculum in science for all students in grades K–12 that contains the following aspects: (a) All the content standards are included and embedded in a variety of curriculum patterns that are developmentally appropriate, interesting, and

relevant to students' lives. (b) Inquiry is emphasized as a tool for learning science as well as a form of content. (c) The science curriculum is connected to other school subjects.

3. Coordinating science with the mathematics program to enhance student use and understanding of mathematics in the study of science and to improve student understanding of mathematics overall.
4. Providing students with appropriate and sufficient resources, including time, materials and equipment, space, teachers, and community.
5. Providing equitable access of all students to opportunities to achieve the National Science Education Standards.
6. Creating schools as communities that encourage, support, and sustain teachers as they implement an effective science program.

Most importantly, however, is the broader view of science content itself. The eight features of content envisioned for schools in the *Standards* include science as inquiry, physical science, life science, earth and space science, science and technology, science in personal and social perspectives, history and nature of science, and unifying and using science concepts and processes. The eight features of the curriculum must not exist in a stand-alone fashion, because they cannot make a coherent curriculum as separate parts. The eighth feature really provides the connection between the other seven features. It is central to an integrated-science approach. It is through using science concepts and process skills that illustrates student understanding and provides evidence for real learning.

PRE-SERVICE SCIENCE EDUCATION AT THE UNIVERSITY OF IOWA: A NATURAL ALIGNMENT WITH SCHOOL SCIENCE REFORMS

Producing teachers with the skills required by integrated science and those included in the *National Science Education Standards* is a necessity if new teachers are not to be out of date before they begin. All efforts in teacher education should include a broader view of assessment strategies and science content. These efforts provide the primary challenges to science teacher education as we re-think and restructure science teacher education. The Iowa pre-service science education program is naturally aligned in the philosophical approaches referenced in the *Standards*; they are also aligned with the features described in three major research studies funded by the U.S. Department of Education and headquartered at the University of Iowa (Salish I Research Project, 1997a, 1997b; Salish II Research Project, 1998; Tillotson, 1995). The pre-service science education program, through a sequence of three methods courses and the student teaching internship, supports the reform tenets and what we know about learning (NRC, 1996). The Iowa pre-service program advocates a student-centered classroom where the teacher is a facilitator rather than a dispenser of information. The Iowa program emphasizes student engagement, the teaching of concrete concepts before abstract concepts, and the relevance of content. To enhance this relevance, the science portrayed emphasizes science that is local and of a personal nature while also illustrating an integration within the sciences and science with other disciplines.

The Iowa integrated-science approach incorporates assessment of science learning and understanding that aligns with student learning based on inquiry, problem solving, and collaboration. Assessment is ongoing, embedded in instruction, and aligned with a constructivist approach to science learning. The *National Standards* provide examples of assessment that characterize the goals and practices found in integrated-science classrooms and in the Iowa Teacher Education Program. (Examples of specific assessment protocols are provided in the Iowa Assessment Handbook [Enger & Yager, 1998].) The Salish I and II research studies (1997a, 1998) provide a basis for defining teacher education successes as well as research instruments for use in similar teacher education practices in other settings (Salish I Research Project, 1997b). Some insight into the nature of the pre-service program and its alignment with a new integrated approach is provided by case studies completed by graduate supervisors involved with the program each semester. Such case studies indicate successes and problems as new science teachers are prepared in Iowa.

At the University of Iowa, changes have been made to the science teacher preparation program that result in new teachers who are reform-minded and who possess the necessary skills to implement reforms. Specific features describe the Iowa Science Teacher Education Program that has been designated as a model since receiving NSF funding over a ten-year period during the 1970's, twenty years prior to the integrated science approach developed during the period from 1990 to 1997. The primary features of the Iowa program that have existed since the 1970's include:

1. A strong science major with a focus on integration of life, physical, and/or earth/space science;
2. Three consecutive methods courses over a three-semester sequence each with a school practicum;
3. A six semester-hour block specifically addressing the nature, history, and sociology of science;
4. A six semester-hour block focusing upon the use of science concepts and skills in resolving current problems;
5. Three practica in schools focusing on problems in elementary, middle, and high schools prior to a semester-long internship (i.e., student teaching);
6. A research project based on a personal rationale statement for teaching;
7. Videotaping and journaling which requires self-analysis of classroom teaching of each pre-service teacher over the course of the methods sequence and the student teaching internship (thereby requiring at least four full semesters with the professional science education staff);
8. Student teaching as fully licensed teachers in integrated science schools; the pre-service are thereby involved in the reform effort as full staff partners.

The focus upon integrated science as reform and the involvement of all student teachers in reform-minded schools is a major step in preparing new teachers for classrooms envisioned in NSES. Since these changes began fully with the NSTA reforms (integrated science), all our experience in Iowa is limited to the 25 students each year who have completed the program (including student teaching) since 1995. A look at the thinking and reflections of a sample of these teachers may provide glimpses of successes and weaknesses of the Iowa model as it currently exists. The program is altered each year as new information and experiences with new graduates are analyzed. Such study provides us an opportunity to design specific modules for preparing integrated-science teachers and for assisting in the development of rubrics to assess the effectiveness of the modules as they are implemented in an integrated fashion.

Classroom visits, extensive interviews, and a study of videotapes of classroom interactions provide evidence of the effectiveness of our efforts in Iowa to prepare integrated science teachers. We focus upon evidence of success with teaching practices that are oriented by a constructivist epistemology throughout the entire preparatory program (science, science education, and general education). This evidence includes determining the degree to which teachers:

1. Seek out and use student questions and ideas to guide lessons and whole instructional units;
2. Accept and encourage student initiation of ideas;
3. Promote student leadership, collaboration, location of information, and taking actions as a result of the learning process;
4. Use student thinking, experiences, and interests to drive lessons (even if this means altering teacher's plans);
5. Encourage the use of alternative sources for information both from written materials and experts;
6. Use open-ended questions and encourage students to elaborate on their questions and their responses;
7. Encourage students to suggest causes for events and situations; and encourage them to predict consequences;
8. Encourage students to test their own ideas, i.e., answer their questions, encourage them to propose their own ideas as to causes, and predict consequences of their ideas and experiments;
9. Seek out student ideas before presenting teacher ideas or before studying ideas from textbooks or other sources;
10. Encourage students to challenge each other's conceptualizations and ideas;
11. Use cooperative learning strategies that emphasize collaboration, respect individuality, and use division of labor tactics;
12. Allow adequate time for reflection and analysis;
13. Respect and use all ideas that students generate; and
14. Encourage self-analysis, collection of real evidence to support ideas, and reformulation of ideas in light of new experiences and evidence.

We look for evidence of understanding and models for assisting in the developing student understanding of the meaning and history of science. We look for evidence that both science and technology are the subjects of inquiry and study. We focus upon question generation and use. Just as the use of science skills and concepts are important, we look for evidence that student engagement is achieved and maintained and that real learning is in evidence. This is of concern in all facets of instruction, both in education and science. We consider the following actions characteristic of a teacher who understands and uses constructivist principles (Brooks & Brooks, 1993; Lutz, 1996; Yager, 2000). The teacher:

1. Encourages and accepts student autonomy, initiation, and leadership;
2. Allows student thinking to drive lessons: Shifts content and instructional strategy based on student responses;
3. Asks students to elaborate on their responses;
4. Allows wait time after asking questions;
5. Encourages students to interact with each other and with you;
6. Asks thoughtful, open-ended questions;
7. Encourages students to reflect on experiences and predict future outcomes;
8. Asks students to articulate their theories about concepts before presenting your understanding of the concepts;

9. Looks for students' alternative conceptions and design lessons to address any misconceptions.

The critical aspects of the Iowa program are the three-semester sequence of methods courses—each with a significant practicum in an elementary school, a middle school, and in high school. The methods sequence consists of a cohort of 20 in Methods 1 that is reduced to 15 for student teaching in the fourth semester of the series. About half of the students are Masters of Arts in Teaching students—the other half are undergraduates with a major in integrated science—but with a concentration in life, earth, chemistry, or physics. All of the teaching traits as well as the integrated science courses and materials exemplify the philosophy and practice previously described.

The sequence of the three methods courses and associated school practica certainly provide the time and the teaching model we hope to develop in each new teacher. There are content strands that are apparent in each of the courses over the three semesters. Inquiry is one of the strands. In Methods 1, students all participate in an inquiry in science and in science education. All students are expected to create at least one inquiry lesson and use it in the elementary school to which they are assigned. In Methods 2, students consider how inquiry can be used and how it frames all that is done. They examine texts and manuals, examine suggested activities, and restructure them to be more inquiry-oriented. The students create an inquiry lesson, use it with middle school students, evaluate their performance, and restructure it for a second use with new students. Methods 3 students are in a secondary school classroom for two periods each day. They spend much more time developing inquiry lessons, changing cookbook lessons into more interesting ones, and planning new lessons as the need for testing student ideas emerge.

Lesson planning is a central activity—again becoming more intense and for longer periods of time during Methods 1, 2, and 3. (On decreasing the resistance to lesson planning see Fetters and Vellom, this volume.) Teaching behaviors provide another strand through the methods sequence. Videotaping of each student occurs frequently. The tapes are used as teaching behaviors are analyzed and tried anew. Examples of instructional concerns include questioning strategies, wait time, classroom management skills, and non-verbal behavior.

The development of a research-based framework or vision statement is another strand. In Methods 1 this is simply a vision statement arising from extensive reading. The research-based framework is a major product of Methods 2. It is rewritten and becomes a part of the growing portfolio for the methods sequence. It is rewritten and polished during Methods 3.

Assessment is a strand that occurs throughout the sequence—almost with all other facets and each day of a methods classroom. Cooperative learning is a feature of the instruction. Its use is also formally examined and evaluated. The research base for instruction is related to an historical view of learning theories. Currently constructivism in all its forms is central to the instruction. Similarly, the application of science concepts and skills to everyday life is featured. The teachers-to-be learn that information that is not connected to living rarely becomes knowledge—i.e., information and skills students can use on their own after leaving the classroom and the teacher.

Other important facets of the program are the application courses that continue to be focused around biology, chemistry, physics, and earth science (mainly to strengthen licensure requirements). These courses are all project-based where students identify questions and issues, propose possible answers or solutions, choose one to collect evidence of its validity, amass evidence of its accuracy and use in taking action, and communicate results and argue issue resolution as an important experience with good citizenship.

We are developing a cadre of faculty members in the science departments who are also integrated-science enthusiasts. New courses are created each year as better examples of the kind of school science teaching are sought and modeled by the preservice students.

A final aspect of the program is the preservice teachers' work related to the history and sociology of science. During this yearlong effort students are asked to examine the total science enterprise, to analyze its procedures, advances, and debates. The teachers-to-be become observers as well as "doers" of science. They learn to question nature as well as science teaching as they become more grounded in the inquiry process. They are encouraged to mount numerous action research projects and to share their actions and the results of these with their peers.

There is no one program—or one series of courses—which fits all students. Instead each student is encouraged to develop his/her own program as he/she follows up on personal questions, uses unique past experiences, is moved to take action to resolve problems and issues. All students are encouraged to reflect on their own actions and to share their thinking, experimentation, and personal actions. The analysis of these situations and contexts experienced by the pre-service students are used to study and to practice good teaching as they interact more with students in schools where they are enrolled for practicum for three full semesters and full-time student teaching for a semester. We also attempt to follow students and to encourage more dialogue, research, and conversations during the first five years of actual teaching.

Following are two stories from our research that documents what the teachers we trained were doing once in the day-to-day practice of schools. These stories exemplify the kind of teacher prepared for the integrated-science approach advocated for all new science teachers at the University of Iowa. Steve and Terry are fairly typical students who reflect on their experiences during a semester long student teaching internship.

INTEGRATED SCIENCE IN THE MIDDLE SCHOOL

Steve was teaching in a middle school with students largely from lower socioeconomic circumstances. He brought to the classroom a great deal of creativity in his instruction and encouraged student engagement by allowing students the freedom to make many of the decisions about the classroom activities. He wanted to develop students' intrinsic motivations by allowing them the opportunity to decide on what they were most interested in, how they would draw out the information they needed, and then allowing them to focus on those areas. Ultimately, he wanted students to develop questions for investigation and to design strategies for answering those questions.

After spending some time in class with the projects designed by his anchor teacher, he realized that the students in his classes were not accustomed to or equipped for the degree of decision-making that he had anticipated. As a result of several days of frustration, he decided to approach his teaching with a little more structure, hoping to teach the students to be more effective in their decision-making. He thought that with practice, they would develop greater autonomy. He described how he engaged his students, gave some structure to their decision-making, and allowed a level of decision-making they could effectively handle in his rationale for his first module:

> When I set out to do a module in the seventh-grade science room I was student teaching in, I had one primary goal: to implement a module that the students would be intrinsically interested in and that would require them to use the science that they learned to explain something that was part of their everyday lives—like a movie. The first time I saw *Jurassic Park,* I knew that it was such a movie. I did not force the module on them. I included it on a list of about 20 topics for them to choose from; however, they, not unsurprisingly, chose *Jurassic Park* as their number one choice.

Once the focus of study was decided upon, students began to investigate a variety of questions that would ultimately lead to constructing *Jurassic Park* in their classroom. He generated questions such as "What would we need to know to build a park of this nature?" "What do we need this knowledge for?" "Why is it important to building *Jurassic Park*?" and "How can science help to make such a park safe for us all?" Other questions led students through some of the typical content of physical science (heat and pressure), earth science (earthquakes, volcanoes, glaciation, erosion, fossil formation, continental drift, climate, and weather), and life science (animal behavior, genetics, DNA, inheritance, and reproduction).

As students engaged more deeply in the study, the questions they generated would sometimes lead them on side trips that extended a single question into series of questions. At this point, Steve felt the students were ready for greater autonomy in the decision-making process and encouraged the side ventures as extensions to their learning. Steve reflected on one extension that led to a whole study in itself:

> Extensions came up now and then and I made sure that we made full use of them. The entire animal behavior was student-generated—entirely! I had asked them to do a research project earlier that dealt with dinosaurs. The goal was to learn about dinosaurs that were not popular or well known. During this, I heard several students comment that they wished they knew more about behavior in order to understand more about their dinosaurs. I talked this over with Claire (my anchor teacher), and we decided to ask the class whether they wanted to take a side trip to look at animals today. They did and I'm glad we did this because I learned a lot. I also got a chance to have the students apply this back to their dinosaurs which helped me assess what they were learning.

Steve used the *Jurassic Park* module to address questions of the current technological capabilities that science offers modern society. He allowed students to debate ethical questions of how such technologies challenge modern society. In addition to those issues, students debated and discussed the moral responsibility of the scientists as they examined scientific decisions from a historical and hypothetical perspective. He also challenged students to examine the nature of science in solving problems and pursuing answers.

Throughout the unit Steve provided opportunities for students to use their conceptual knowledge, their problem solving abilities, and their creativity through a variety of forms of assessment. Steve's discussion of his assessment of *Jurassic Park* demonstrates his desire to meet the individual needs of his students while providing an instrument that reflected the goals and objectives of student teaching. He said:

> First, I wanted to make sure that each student had a chance to perform well on a quiz well suited to his or her learning style. Second, I wanted to see what I liked and what I didn't. The one quiz format I used several times asked the students to take the science they had learned in the present tense (e.g., animal behavior—at one point we talked only about animals living today) and apply it back to the dinosaurs in a hypothetical sense. This allowed me to see not only that the students were able to generalize, but that they were also able to do it abstractly and creatively. This also had them thinking more like scientists (nature of science goals). The students responded favorably to this, although some found it hard to do. With practice, it did get better. Furthermore, I decided that the best assessment at the end of the module was to watch the movie. Then, at various points the students would be able to use the science they had learned to explain a part of the movie. I would stop the movie and give them time to explain what was going on using what they had learned.

Students determined that they needed an outlet to share what they were finding out and what was happening in their class. When their classroom resembled *Jurassic Park* as closely as they thought they would be able to achieve, they issued an invitation to the first grade classes from the elementary school. A group of three seventh graders acted as tour guides for as many as ten first graders and presented information and explanations of the history and science behind each part of their "park". Steve reports:

> This was probably the most rewarding part of *Jurassic Park* for the students and the teachers. The little kids were very excited to be here, see what we were doing, and having the "big kids" teach them. In addition, they were also studying dinosaurs at this time; therefore, our *Jurassic Park* went well with their curriculum.

The seventh graders planned a variety of "hands-on" experiences that would allow the first graders to get involved in some of the same activities they had done during the module. One of the first grade teachers commented to Steve's college supervisor, "We don't get many opportunities to take field trips because of the cost involved in trips. For me, this is the best field trip we could have taken because our students got a chance to reinforce their own learning and also have the opportunity to see older children excited about learning. For us this is a real treat!"

In his final reflection on the *Jurassic Park* module, Steve looked back at the goals he had for *his* learning experience. He wanted to use this experience to improve his instruction in three areas. First, he wanted to learn more about what it takes to become an effective facilitator in the constructivist learning experience. Second, he wanted to include both within-discipline perspective as well as an interdisciplinary perspective. Finally, he wanted to refine the use of cooperative learning, an instructional strategy to which he had been introduced during one of his practicum experiences.

To become effective as a facilitator, Steve researched intrinsic motivation and learned the value of providing choices as a means of motivating individuals. "Not only do choices motivate but they also provide opportunities for students to become excited about learning. It just makes learning more meaningful". The difficulty he faced during his teaching experience, however, was that students had been provided so few opportunities to make choices in their previous classroom experience that they had to be given some guidance about how to make effective decisions that led to learning. This frustrated Steve and, when asked what he would do differently, his immediate response was:

> I believe I would give the students even more choices over what they wanted to learn about the project overall. I did this some, but not as much as I would have liked, since the students were not used to this. I felt that giving too much freedom at this stage would actually have been more of a hindrance. I feel that this comes with the students knowing you and trusting you and each other. We have achieved this for the most part now, and in our next module I am giving the students almost complete control over what topics they want to study, how and when, as well as how they wish to be assessed.

The *Jurassic Park* module provided Steve with an opportunity to develop relationships with other disciplines within science while learning in the science classroom. Steve incorporated mathematics (measurement relationships and comparisons), English (creative writing, storytelling, essays to explain and defend positions and concepts), art (design and construction of models and diagrams), history (relationships of past and present), social studies (archeological studies), psychology (stimulus/response reactions and behaviors of animals), economics (costs of the unit), and ethics (morality and ethical decision-making in science and society).

Steve reached his third goal ("to spend the semester working with, studying, and learning more about cooperative learning") by organizing the module around cooperative groups of three to five students. He reflected:

> All of the student projects were based on different aspects of cooperative learning as set out by Dr. Stanley Kagan. This was a challenge for the students as they hadn't used anything else like this in their classes, and had not experienced use of cooperative groups since fifth grade. They needed to be "retrained" on how to work cooperatively and what the goals of such learning were (and that sitting together did not mean simply talking about anything they wanted). I also made sure that every activity required input from each member of the team to make sure that one student was not carrying the group.

Moving into uncharted waters (full responsibility for reaching and implementing current reform strategies), Steve found that interaction and communication with both administrators and parents were vital components to the success of his instruction.

> This is important to me, especially since what I was doing was untested, and I am only a student teacher. Therefore, I talked it over with the principal before I started, and she was in favor. I also sent home explanations to all the parents to let them know what their kids would be studying. Finally, I explained by theory the rationale behind *Jurassic Park* to all the partners who came to see me at conferences (this was about half the seventh grade). I was pleased to see that not only were they interested, but they were also glad to see something different going on and they all had good things to say about science and what I was attempting. This made me feel good. They told me how their

children were coming home and talking about what was going on in science—for the first time in years.

INTEGRATED SCIENCE IN THE HIGH SCHOOL

During a sequence of three science methods courses at the University of Iowa, science education students write and revise a rationale for their science teaching. Practicum experiences, which are a component of all the methods classes, provide opportunities for the prospective science teachers to test and revise their rationales based upon field experience. Teaching rationales are grounded in theory and research but are also revisited from an experience base. The view of Terry presented here is derived from his rationale paper, student-teaching journal, student-teaching videos, analyses of questioning skills, anchor-teacher evaluations, and finally university-supervisor evaluations and conversations.

Refining a Rationale for Science Teaching

From the beginning, Terry conveyed the desire to have students become not only scientifically literate, but he also considered that for science to have relevance it must be applicable to the daily lives of students:

> When I start my career as a high school science teacher, I realize that it is important to set certain goals for my classroom. I want my students to gain an appreciation for science and learn to apply scientific knowledge in their everyday lives.

Based upon his practicum and third methods course, Terry felt:

> After completing Methods III and my practicum, I have reviewed my rationale and am satisfied with my original goals and methods. Through my practical experience, I have a more concrete vision of how to use the methods in my original rationale and new ideas on how to achieve my goals. My experience has given me great insight into the significance of classroom management and many of the ways to have effective classroom management.

Terry also indicated that he planned to use cooperative learning and wanted to foster an environment in which classroom science activities were connected with the interests of his students. Fostering the creativity of students through the use of student-centered activities was of importance to him. Terry intended to include in his teaching daily feedback to students with respect to their work and frequently use quizzes to check for student understanding. He planned to assess in ways that required application of science problem-solving skills to real world contexts so that students would apply their knowledge to everyday life instead of just memorizing for a test.

Terry stated in his consideration of classroom management that he would use a proactive style and hoped to offset potential misbehavior by maintaining student interest. He planned to avoid power struggles by minimizing confrontations with students in front of the entire classroom and planned to manage students in a more private teacher-student dialogue.

Evidence From the Student Teacher Experience

Terry's primary student teaching assignment was in a secondary physical science classroom. He had a strong academic preparation in physics and engineering as well as in chemistry. Terry's cooperating teacher held a constructivist philosophy that matched the student-centered approach outlined in Terry's rationale paper. This provided a natural alignment with the philosophy and tenets of an integrated science classroom.

Terry began working in the physical science and physics of technology classes on his second day at school by discussing homework assignments with the students. His cooperating teachers also involved Terry in setting up laboratories for the classes on the second day. Terry discovered very early on that preparation is a key to being personally at ease in the classroom. The amount and quality of equipment available for use in the classrooms surprised Terry. These resources were used when facilitating hands-on work with students. In a laboratory in which students were setting up electrical circuits, Terry noted that the experience provided him as well as the students with problem-solving situations, which were not prescriptions but ways to work through a learning experience. Computer applications were also a part of the physics classroom.

As the semester progressed, the students continued in their work towards open-ended laboratory formats. The time required to plan lessons and provide feedback to students was substantial; it was very apparent that a teacher's day extends both before and beyond the scheduled hours ("I stayed until about 5:30 to grade Monday's lab. I took two sets of essays home to grade").

A point of frustration for Terry was students' failure to work at problems beforehand. If the answers were not easy to derive, students did not try to do homework.

> Many of the students were having tremendous difficulty with working the physics problems that were assigned. Many of them had not really tried the problems. I hope to work with them so that they realize that you cannot expect to know how to solve a problem as soon as you are done reading it.

Terry hoped to improve the students' problem-solving skills through problem-solving interventions. To help facilitate students in developing their problem-solving skills, Terry drew from a research base:

> I mentioned to students that research has shown that people learn material more thoroughly when they have to explain it to someone else. Based upon this idea, I would like you to work in self-determined groups of three and work at solving problems together and communicate your understanding to others in your group. I will also be here to work with you.

Based on his observation of work in cooperative groups, Terry indicated that students did make progress in improving their problem-solving skills. The lack of student preparation for quizzes and examinations and student expectations to have a retest when they did poorly also bothered Terry ("When I went to school, we never had that opportunity").

Laboratory work was used extensively to engage students in both hands-on and minds-on activities. Terry implemented not only more traditional physics laboratories but also worked at facilitating a more open-ended laboratory approach. Terry described some aspects of laboratory work that raised his level of concern:

> The students did a lab where they observed the images formed by a spherical concave mirror. It was difficult for students to make good observations. I circulated around to help students and attempted to help students to think about what they were asking instead of giving them answers. Students seemed to be almost trained to expect the teacher to answer a question without directing the question back to the students to have them attempt to answer the question for themselves. I want my students to develop more independent thinking.

Terry had stated ideas about what he planned to do to assess student learning. The students in physics were provided with an assessment project in which they were to design a product that was aligned with concepts from an optics unit. A rubric with criteria was provided and discussed with students who would work as partners on the assessment. Terry made these observations:

> Many of the students are trying unique ideas such as periscopes that also magnify. Some also attempted more difficult ideas such as concave mirror telescopes. One problem is the need to share supplies across classes. I do think that many of the students were enjoying the challenge and less structured nature of this assessment project.

The experience in various classes was also a source of insight into the world of the classroom.

> I really enjoy teaching physics of technology classes. When students come to class, many of them are interested in what is going on with the physics course and often I show them the demonstrations that I do for physics. They are often more interested than the physics students. For this reason, I relax a bit more in this class and am not as strict, which usually does not pose any problems. However, sometimes these students do push the limits; but if I get a little upset, then they usually back down. I find this humorous because I know they cause problems for many teachers.

At the University of Iowa, we require student teachers to have experience in a variety of classrooms as a part of the model described earlier. When Terry began teaching in a ninth-grade integrated-science classroom, students were working in teams on a project to design an organism that could survive on another planet. To help students consider factors such as gravitation, he used a hands-on approach.

A rubric for the assessment of a culminating product, an organism that was adapted to another planetary environment, was shared with the students so that they were familiar with the criteria on which the product would be evaluated. Terry found the students developed some very creative organisms.

> The students have been working on alien designs and some of them have some very creative ideas and have constructed a visual of their alien. The students got together with other people in the class who had designed aliens for the same planet and determined how they were going to interact with each other. It was cool to listen to them figure out how their aliens would interact. The students also organized their class presentations.

Class sizes, which averaged about 28 students, also presented some management challenges not experienced in the physics classroom. He asked "students to sit quietly for a moment in complete silence," which helped to calm the situation. While working in an astronomy class, Terry again noted the problem of student engagement ("Many of the students seem like unmotivated students who are not really too interested in learning astronomy or anything else for that matter").

Insights from Video Tapes

Videotaping of whole-class periods was performed frequently—usually with the cooperating teacher at the camera. Good taping usually recorded student-student interactions as well as those involving the teacher with individual students. A series of lessons were recorded each month of the semester-long student-teaching experience.

Videotapes provided important feedback that allowed Terry to view himself in the role of classroom facilitator. The questioning levels could be focused upon as well as a close up look at the image he portrayed as a teacher in the classroom. The videotapes provided a setting in which he could judge and reflect upon his teaching practice. The videotapes provided an opportunity for Terry to examine what he did well and also allowed him to assess ways in which he could polish his classroom presence. Videotapes are also viewed and critiqued by the university supervisor often using the ESTEEM model (Burry-Stock & Oxford, 1993) and involving other student teachers. The following comments were in Terry's written documentation from self-analysis of such tapes:

> I say "OK" a few times in the beginning. I say "uhm" too much. I need to use more inflection in my voice. I need to keep students on task. I am more interactive than I was on the first tape. I used good wait-time. Instead of asking students if they have any questions, I ask what questions they have for me. I strive in all my classes to help the students become more independent of me as the teacher and to not rely on the teacher to bail them out every time they are challenged.

Insights from the Student Teacher Supervisor

Terry found the student teaching experience to be most educational because theory and ideas could be placed into practice. (On the relationship between theory and practice, see also the chapter by Roth.) In the reality of the teaching experience, Terry found that what is held as the ideal classroom is very quickly challenged by the reality of the actual classroom. Trying to foster a learning environment that is based on research becomes difficult when students are apathetic. Attempts to correct this apathy do not work immediately, if they work at all.

The supervisory role was approached from a mentorship position. This direction promoted a forum for constructive dialogue and interchange of ideas ranging from designing activities, classroom management strategies, and friendly professional conversations. Yet, because he did not participate in teaching as described by Roth or provide the kind of modeling described by Thompson and Hargrave, there remained a difference between the two with respect to the object of analysis.

While constructivism was the underpinning philosophy in the science classroom, the classroom teacher and district curriculum guidelines did predetermine the curriculum agenda to a certain extent. The cooperating teacher, however, was flexible in allowing Terry to try out ideas. Terry's cooperating teacher demonstrated his confidence in Terry's academic preparation and management skills by involving Terry in a leadership role very early in the student teaching experience.

Insights from the Cooperating Teacher

Terry's cooperating teacher and primary mentor was an experienced professional whose guidance was very helpful to Terry. Communication between Terry and his mentor was enhanced because of a shared vision about what science education should be and thinking about how students learn. In conversations with Terry's mentor, it was clear that Terry's academic training, methods courses, role models, and experiences from other employment were all useful in Terry's move into a successful student teaching experience. Rough edges did exist in areas such as being able to plan lessons for both short term and long range. Classroom management skills also evolved and developed under the mentorship. None of these were viewed as major problems but areas requiring continued attention and more experience. Terry established an excellent working relationship and rapport with his students. He was viewed as a fine new teacher who would be welcomed as a regular teacher in the school where he did his student teaching.

SUCCESS IN PREPARING TEACHERS TO TEACH INTEGRATED SCIENCE

Steve and Terry are but two of the teachers who recently completed the University of Iowa science teacher pre-service program. Both case studies showed that the individuals became increasingly reflective about their teaching in the way that our program was designed to foster teacher development. The experiences of Steve and Terry are rather typical for the graduates from our program. With a focus upon Project 2061 (Rutherford & Ahlgren, 1989), Scope, Sequence and Coordination (NSTA, 1992; Pearsall, 1992), and the National Science Education Standards (NRC, 1996), it is to be expected that our graduates are knowledgeable of current reforms and anxious to be involved. At the same time there are constraints in terms of skills and school traditions and philosophy for implementing reform ideas. The case stories illustrate the strength of the Iowa program while also indicating the problems with implementing new ideas with teachers-in-training. Frequently, the high school program remains discipline bound by course titles. However, a more integrated science approach is needed as students relate their study to their own lives and communities. Such concerns mean less reliance on textbooks and to a set curriculum framework.

Student teaching constitutes the capstone experience in which the student teacher has an opportunity to implement their pre-service training. The student-teaching experience is enhanced when the mentor teacher shares common philosophical beliefs about how students learn science and how science instruction can facilitate student learning. Placement of a student teacher in a classroom where an integrated science approach to instruction and learning is being practiced provides a model that is a powerful agent in changing the nature of science classrooms and nature of the science encouraged by new science teachers. (See Roth, this volume, on the mutual benefits that arises for all partners in coteaching.)

The stories of Steve and Terry are not unique. The Salish I and Salish II research efforts involved 30 other Iowa graduates. Studies by Tillotson (1996), Craven (1997), and Hollenbeck (1999) clearly indicate that open classrooms tried by Steve with *Jurassic Park* and the open discussions and laboratories Terry provided are common. The Salish I research was shocking in illustrating that much of the idealism disappeared as new teachers found themselves alone—and often without caring mentors which the student teaching experience provided. Nor were university supervisors available or involved. The consequence of these effects is considerable as the following paragraphs show.

The science teachers that we prepared for an integrated-science approach in secondary schools often take positions in high schools in which this integration is considered to be radical. Working in such contexts encourages these new teachers to revert to more teacher-directed instruction and more textbook-dominated content. We can easily imagine that a longer experience in apprenticeship (see Thompson and Hargrave, this volume) or coteaching (see Roth, this volume) situations would stabilize the initial practices to the point that new teachers were able to maintain them even in adverse conditions. The good news is that nearly all teachers regain confidence and report the validity of their philosophies developed over a two-year period of preparation of the University by year three of full-time teaching. This becomes much more pronounced by Year 5 in teachers' careers (Hollenbeck, 1999). The problem still unresolved is the retention in classrooms for five years and beyond. Both Steve and Terry have since left teaching. It seems that issues concerning induction into teaching need more attention and resolution. These programs must move beyond having a mentor teacher assigned. And when assigned, the mentors need to reinforce the importance of more active involvement of students in determining instruction and content. Both must be aligned to the new visions included in the national standards (NRC, 1996). As mentioned, apprenticeship and coteaching models might be a solution to our problems.

Sternes and Paris (2000) illustrate the necessity of students choosing to learn in order to learn. So it is with new teachers. They must be familiar with the research that has lead to new visions and they must choose to change their instruction and a more integrated approach to science or they will not be effective in their chosen profession. Both Steve and Terry had doubts; both were reflective about their teaching; both reevaluated their rationale for teaching; both proved ready to help with current reforms. It was interesting to follow them into teaching—Steve for three years; Terry for four. Both needed a better system for assuring continued growth and success with teaching an integrated-science approach.

Developing new teachers to move school programs to integrated science requires time. Three semesters with methods each coordinated with school experiences, a full semester of student teaching with an anchor teacher and perhaps three other cooperating teachers is important—but not adequate to produce new teachers able to succeed alone. There is evidence that a larger and well-planned induction program is essential. Perhaps such a program will assure that teachers like Steve and Terry will remain in the classroom as teachers continuing to learn in a supportive environment for an entire professional lifetime.

REFERENCES

Brooks, J. G., & Brooks, M. G. (1993). In search of understanding: The case for constructivist classrooms. In R. E. Yager (Ed.), *Science/technology/ society as reform in science education* (pp. 59-67). Albany: State University of New York Press.

Burry-Stock, J. A., & Oxford, R. L. (1993). *Expert science teaching education evaluation model (ESTEEM) for measuring excellence in science teaching for professional development.* Western Michigan University.

Craven, J. A. III. (1997). *Relationships between new science teachers' beliefs and students perceptions of the learning environment.* Unpublished doctoral dissertation. University of Iowa, Iowa City, IA.

Enger, S. K., & Yager, R. E. (1998). *The Iowa assessment handbook.* Iowa City: University of Iowa Science Education Center.

Hollenbeck, J. (1999). *A five year study of the attitudes, perceptions and philosophies of five secondary sciene education teachers prepared in the constructivist teaching methodology advanced at the University of Iowa.* Unpublished doctoral dissertation, University of Iowa, Iowa City, IA.

Kimble, L. (1999). *A comparison of observed teaching practices with teacher perceptions of their teaching during and following major funding.* Unpublished doctoral dissertation. University of Iowa, Iowa City, IA.

Krajcik, J. (1986). *An evaluation of the university of Iowa's science teacher education program, 1977-1984.* Unpublished doctoral dissertation, University of Iowa, Iowa City, IA.

Lochhead, J., & Yager, R. E. (1996). Is science sinking in a sea of knowledge? A theory of conceptual drift. In R. E. Yager (Ed.), *Science/technology/society as reform in science education* (pp. 25-38). Albany: State University of New York Press.

Lutz, M. V. (1996). The congruency of the STS approach and constructivism. In R. E. Yager (Ed.), *Science/Technology/Society as reform in science education* (pp. 39-49). Albany: State University of New York Press.

Mestre, J. P., & Lochhead, J. (1990). *Academic preparation in science: Teaching for transition from high school to college.* New York: College Entrance Examination Board.

National Research Council. (1996). *National science education standards.* Washington, DC: National Academy Press.

National Science Teachers Association. (1992). *Scope, sequence, and coordination of secondary school science: Vol. I The content core. A guide for curriculum designers.* Washington, DC: Author.

Pearsall, M. K. (Ed.). (1992). *Scope, sequence, and coordination of secondary school science: Vol. II Relevant research.* Washington, DC: National Science Teachers Association.

Perrone, V. (1994). How to engage students in learning. *Educational Leadership, 51*(5), 11-13.

Rutherford, F. J., & Ahlgren, A. (1989). *Science for all Americans.* New York: Oxford University Press.

Salish I Research Project. (1997a). *Secondary science and mathematics teacher preparation programs: influences on new teachers and their students.* Iowa City: University of Iowa, Science Education Center.

Salish I Research Project. (1997b). *Secondary science and mathematics teacher preparation programs: influences on new teachers and their students—instrument package and user's guide: a supplement to the final report of the salish i research project.* Iowa City: University of Iowa, Science Education Center.

Salish II Research Project. (1998). *Translating and using research for improving teacher education in science and mathematics.* Iowa City: University of Iowa Science Education Center.

Scott, G. (2000). Integrated science study. *The Science Teacher, 67*(6), 56-59.

Stiles, J. (1993). *A study of a science education preservice program model and its effects on the attitudes, perceptions, objectives and philosophies of its students.* Unpublished doctoral dissertation. University of Iowa, Iowa City, IA.

Tillotson, J. (1996). *A study of the links between science teacher preparation program features and new teacher performance with regard to constructivist teaching.* Unpublished doctoral dissertation, University of Iowa, Iowa City, IA.

Yager, R. E. (1995). The constructivist learning model: Toward real reform in science education. *The Science Teacher, 58*(6), 52-57.

Yager, R. E. (2000). The constructivist learning model. *Science Teacher, 67*(1), 44-45.

Yager, R. E., & Lutz, M. V. (1994). Integrated science: The importance of "how" versus "what". *School Science and Mathematics, 94*, 338-346.

CRITICAL MULTICULTURALISM AND SCIENCE TEACHER EDUCATION PROGRAMS

Norman Thomson, Margaret Wilder, & Mary Monroe Atwater
University of Georgia

INTRODUCTION

In this chapter, we examine secondary-science teacher-preparation programs located in three universities. We focus on how each program has developed and implemented multicultural education initiatives as part of their preservice teacher preparation. Recent national and statewide calls for improvement in preservice teacher education have spurred college programs to better prepare science teachers to teach *all* students, especially those from underrepresented groups. The primary goal is to ensure that all K–12 students receive high quality experiences in science education.

We presuppose that for multicultural science education to be critical, it has to be attentive to Ogawa's (1998) criteria for critical teaching and learning. Thus, science and science education are "interpretations or constructions by the people of a given culture" (p. 139). In other words, "there is no culture-free interpretation of science or science education" (p. 140). This perspective is crucial for secondary science educators faced with meeting reform initiatives in secondary science teaching. It suggests that science is not independent of learner, teacher, or context. Rather, science teaching and learning can and should include interrelationships between people and environments.

We begin by offering a brief description of educational issues that have generated national attention. Next, critical multiculturalism is defined and briefly discussed in relation to secondary science teaching. We then describe and compare three preservice secondary-science teacher-education programs by focusing on the ways in which these programs attempt to develop critical multicultural teachers. We conclude this chapter with implications for future program development and research.

SIGNIFICANT SECONDARY TEACHER EDUCATION ISSUES

Student composition, school personnel, curricula, and classroom facilities are constantly changing to keep pace with technological and economic changes in the USA. Many public school systems, especially those with fewer resources and located in urban or rural communities, are often overwhelmed with social and financial problems. Even though urban schools are often perceived as serving students of color of low-income status, there is actually a spectrum of urban schools, including White,

D.R. Lavoie and W.-M. Roth (eds.), Models of Science Teacher Preparation, 195–211.

private, and upper-middle-class schools that exist in urban communities (Wilder, 1995). In response to the social and financial problems, private groups now manage some urban schools (Amphrey, 1997; Education Commission of the States, 1997; Green III, 1997). However, even private groups must consider the essential elements of pre-college schooling with which public officials contend. These include student enrollment patterns, school personnel qualifications, instructional methods, curriculum design, financial accountability, school governance, and environmental characteristics and influences. With respect to science education, there have been discrepancies in how science is taught to students in different locales. This is partially due to structural and social inequities that prevent some US schools from providing quality education. This is particularly illustrated in poorly financed public high schools that do not have the laboratories and other necessary physical and human resources to enhance or ensure quality science teaching and learning (Kozol, 1991; National Research Council, 1990).

The mainstream curriculum in US schools has marginalized many students (Banks, 1981, 1988, 1993). Marginalized students, particularly students of color and of low-income status, are seldom represented in the various books and instructional materials used to teach them. Banks (1993) has suggested that teachers face the challenge "to make effective instructional use of the personal and cultural knowledge of students, while at the same time helping them to reach beyond their own cultural boundaries" (p. 8). Rather than discount personal and cultural knowledge, such knowledge should become a vehicle to motivate students and should serve as a foundation for teaching.

Science education teachers have typically taught in homogenous classroom settings (Atwater, 1996a). Many organizations engaged in reform (i.e., American Association for the Advancement of Science, 1989, 1993; National Research Council, 1996) have proposed initiatives to address mono-cultural classrooms and other similar imbalances. The unifying theme that has emerged from national reforms is that all students must become more successful in the science, mathematics, and technology learning enterprise (National Commission on Teaching and America's Future, 1996; National Science Foundation's Directorate for Education and Human Resources, 1997). The common goal of learning for all students is to help secondary science teachers teach so that *all* students can feel a part of the science learning process. To accomplish this goal, policy makers and state funding agencies are encouraged to allocate additional financial support to schools that reside in low-income neighborhoods (NRC, 1996) so that the teachers who teach there have access to the required up-to-date scientific and technological equipment that ensure excellence in science education.

The National Research Council (1996) outlines a framework of standards for science teacher education programs but has failed to address adequately how such goals for all students are to be funded or how they might take shape (Rodriguez, 1997). Moreover, the American Association of Colleges for Teacher Education (Chavez, 1996) and the National Council for Accreditation of Teacher Education (1997) have diversity, equity, and multiculturalism as centerpieces of their teacher education documents. A recent set of standards for science teacher preparation adopted by the National Science Teachers Association (1998) is now included in

NCATE's accreditation process and stresses multiculturalism in some of its indicators. However, these documents do not outline plans of action for teacher educators. Instead, program restructuring has to be funded by the colleges.

Program autonomy is often desired, yet frequently problematic, particularly in matters of funding. Teacher education programs with greater resources such as those found among the larger research universities can fund more sophisticated programs than smaller institutions with fewer resources. Such resource discrepancies at the post-secondary level make it increasingly difficult for all preservice teachers to benefit from various community-school partnerships (Wilder, 1995). Discontinuities among science teacher preparation programs mirror differences found among high schools with fewer resources. Thus preservice teacher-preparation programs and high schools will need adequate funding if high-quality science instruction is to be accessible to all students.

Very little research has been conducted in bridging science teacher education and multicultural education (Atwater, 1996b). However, there is an expanding body of research focused on teacher preparation, teacher practice, and multicultural education (Anderson, 1994; Baptiste & Baptiste, 1979; Gay, 1995; Ladson-Billings, 1994; McLaren, 1989; Page, 1987; Wilder, 1995). There is a broad agreement that learning is a constructive process, involving teachers' prior learning and experiences. The next section presents a brief discussion on critical multicultural education and how it can inform science teacher education.

CRITICAL MULTICULTURAL EDUCATION AND SCIENCE TEACHER EDUCATION

The sciences are couched in their own traditional academic cultures and these cultures have been the domain of elite, White males (Hess, 1998; Jegede, 1994; Ogawa, 1998). Even today, most science education content, pedagogy, and assessment at the post-secondary level are Euro-centric. Although preservice secondary science teachers as individuals develop their sense of self from their home life, as well as from their schooling in science, teacher educators and science education researchers continue to be educated to view and approach science in isolation of self and social phenomena. Generally, knowledge is scientific only if it is perceived as "objective" and "value free" knowledge. Since science education draws upon the natural sciences for content, it too tends to be reductive in its theories and practices. We suggest that preservice science teachers construct and interpret science as it was taught and explained to them prior to entering their science teacher education programs and through informal learning channels such as public media and museums. In addition, most preservice science teachers enter their profession without ever having participated in science as it is practiced (Latour, 1987; Ogawa, 1998).

In most cases, new knowledge is developed through our social interactions and individual pursuits (Ausubel, Novak, & Hanesean, 1968; Vygotsky, 1978). Consequently, preservice teachers' knowledge is also reinforced and learned when they come in contact with their students, their students' parents, the school curriculum, and school officials. Whether and how preservice students are able to construct a more transformative knowledge base for science will depend greatly on how well they manage and accept the interactions that will evolve from these encounters. It will also depend on how well they are prepared to understand and teach the scientific enterprise as a multifaceted, real-world system. When the complexity of science is combined with the complexity of human culture and behavior, critical multicultural learning frameworks are useful in explaining the synergism that emerges from student-teacher interactions. Quality learning is a social activity and should not be confined to or seen merely as an individualistic engagement with the world (Cobb, 1994). The approaches taken in science learning, teaching, inquiry, problem solving, and experimenting depend on the way teachers and students view the world (Atwater, 1996b). Atwater posits that with a critical multicultural lens knowledge of the world, including science, does not exist unquestioned.

Critical multiculturalism is based on beliefs derived from critical theory and multiculturalism. Multicultural education is broadly defined as a research and pedagogical field of study aimed toward achieving the equity of all learners. Multicultural science education focuses on the social, cultural, and political forces that affect the quality of science teaching and learning. Critical theory is a set of beliefs that focuses on the ways that injustice and subjugation shape society (Guba & Lincoln, 1994; Kinchloe, 1993; 1995; McLaren, 1995). In the field of education, critical theorists believe that schools should be institutions where knowledge and values are taught for empowering people to change their worlds rather than a place for educating people for conformity and subjugation (Giroux, 1985; Lather, 1991).

Multiculturalism recognizes that age, class, disabilities, ethnicity, gender, sexual orientation, geographical region, language, place of residence, race, and religion are embodied in institutionalized structures, especially in education (Grant & Ladson-Billings, 1997). Critical multiculturalism is realized through critical analysis of institutions and societal structures and examines conformity, oppression, and subjugation as the results of different cultural groups. It requires "competence in multiple ways of perceiving, evaluating, believing and doing work that demonstrates cultural understanding" (Bennett, 1995, p. 14). Critical multicultural science education encourages preservice secondary science teachers to question and change the power relations that shape and influence science education. The discourse of critical multicultural education can assist science educators in ensuring that all students are included in the learning process. Secondary science teachers should, therefore, use critical multicultural frameworks and incorporate critical multicultural pedagogy in their science courses (Atwater & Riley, 1993).

THREE EXISTING SECONDARY SCIENCE EDUCATION PROGRAMS

The three secondary science education programs discussed here have been selected because they offer a diversity of approaches that share a common goal. (Each university has been given a pseudonym so that their identities will remain anonymous.) They attempt to develop secondary (middle and high school) science teachers who can serve all students. Several factors emerge that reflect and provide useful strategies and components. These might be used to guide and nurture development of multicultural science teachers. In this context, a science-education program is not an isolated entity within a university. Rather, institutional systemic policy reforms provide a forum, impetus, support, and opportunities for change extending from in-house recruitment to off-campus community service. Second, conversation and action between and within university science and education departments are needed to address the dual components required for the development of multicultural science teachers. Both components require specific knowledge of content and processes and the ability to design, implement, and evaluate the learner and meaningful environments in which learning can occur. Third, teacher educators must make long-term commitments, engage in constructive dialogues, and develop programs which, in turn, provide preservice science teachers with opportunities to become exemplary competent role models, reflective practitioners, and life-long learners.

University at Riverside

The University at Riverside is a medium-size institution (12,000 undergraduate, 8,000 graduate students) in the northeastern United States about 100 miles from a major metropolitan area noted for its diverse population. Until recently, the *de facto* role of the School of Education has been cultural reproduction and preparation of science teachers for suburban and rural environments reflecting their origins and professional aspirations for returning "home". Similarly, the science education teacher program's mission and enrollment neither reflected the diversity of the State's or the local population demographics. According to the U.S. Bureau of the Census, the state's estimated population in 1997 is composed of 18% African Americans, 0.3% Asian Americans, 1.5% Hispanics/Latinos, 0.3% Native Americans, and 77% Whites (New York State Data Center, 1998). The local school population is primarily composed of African Americans and Native Americans (New York State Data Center, 1998). In contrast, the university's science education program has been comprised almost exclusively of Whites of western-European origin. A statewide surplus of science teachers has been constantly produced, especially suburban "want to be" teachers, but there exists a perpetual critical shortage of qualified teachers, especially in the physical sciences. Forty percent of the annual available teaching positions are in the state's urban schools and are not considered viable teaching options for many preservice science teachers.

Nonetheless, changes have taken place. A new university president actively pursued the goal of having the university reflect the state's population demographics and developed an infrastructure within the institution that could assist all staff, faculty, and students to meet this goal. For example, mentoring programs, a center for multicultural education, and workshops for faculty, students, and staff now exist as part of the new infrastructure. The School of Education not only responded to the university president's vision, but also developed its own systemic reform initiatives. Its challenge and mission included the preparation of teachers. Thus, it included (a) the identification, recruitment, and education of teachers qualified to successfully teach throughout the state's school system, (b) the development of multicultural courses within all departments reflecting the multifaceted perspectives required for teachers to teach in multicultural environments, and (c) the creation of a certification program of specialization in multicultural education. In part, the latter, was accomplished through the development of at least one multicultural education course being offered in each school's department.

Recruitment of preservice science teachers occurred in two ways. First, to address immediate needs, science teacher education faculty became visible participants in science department colloquia and seminars, and science faculty became involved in science education curriculum reform. Furthermore, one science-teacher educator recruited potential teachers from underrepresented groups in science and science teaching by making presentations at various student organizations such as sororities, fraternities, and the Charles Drew Society. The second, long-term approach was through afternoon on-campus science programs for high-school students to provide opportunities to learn in a university environment and engage in meaningful activities with university faculty. These programs also served to recruit future teachers.

Preservice science teachers were encouraged to participate in the newly formed after-school science programs at all levels in urban schools. One particularly poignant story emerged in a middle school. A group of children who were considered African Americans based on their skin color and who were non-respondent, unmotivated, and passively disruptive became suddenly engaged in an afternoon science program when a preservice Native American teacher was able to interact with them in a meaningful way. Subsequently, the Native American teacher was able to inform the school's administration and faculty members that a different teaching style was required to transform the classroom because the children were Native American children. Once recognized, the children became engaged in all of their subjects. Preservice science-teaching practicum now requires that at least one eight-week block be spent in a public urban school. At least one course in multicultural education is a prerequisite to that experience. In addition, students are encouraged to fulfill the multicultural specialization requirements. Student teachers are expected to develop learning environments relevant to many cultures and experiences. For example, most chemistry texts present the origins of iron smelting in Middle England. However, research articles (Childs & Killick, 1993; Schmidt & Childs, 1995) describe its true origins as African and an article by Murfin (1996) provides relevant laboratory activities.

The transition to a multicultural education perspective has created research opportunities for science faculty, attracted national funding for pre-college programs, undergraduate scholarships, and a better working relationship of the University with local schools through outreach initiatives. Finally, preservice science teachers now feel more qualified and comfortable in pursuing teaching positions in schools districts that do not reflect their "homes". Additionally, local school districts are more comfortable in knowing that the science education program has prepared teachers that are ready to engage with and meet the needs of all their students. These goals and expectations were not possible even a few years ago.

University at Oceanside

The University at Oceanside is a large institution (25,000 undergraduate, 10,000 graduate students) located in the southwestern part of the United States. It is located in a major metropolitan city and is in a state that has been embroiled in issues focused on enculturation of large and diverse immigrant populations. In 1997, the U. S. Bureau of the Census estimated the state's population as follows: 7% African Americans, 11% Asian Americans, 29% Hispanics/Latinos, 1% Native Americans, and 52% Whites (California State Data Center, 1998). At one time, the School of Education was located in an affluent neighborhood but nowadays finds itself adjacent to a vast economically impoverished (though culturally rich) urban area. In recognition of its own role and the dramatic changes in the demographics of the United States classroom, the School of Education has established a center for urban education studies. The Center has a two-fold approach for addressing educational issues found in urban schools: research and teacher preparation. The Center's research foci are to provide frameworks and direction for educational innovation, to investigate the critical roles played by family and community in children's lives, and to document how pressures and changes in these institutions affect teaching and learning. The Center also focuses on issues of school reform, equity, resource allocation, and community development as they are affected by policy proposals such as school district restructuring, decentralization of leadership and accountability, and public choice.

The Center includes a science education program founded on social-justice principles and commitments to integrate theory and practice. The program prepares teachers to have the commitment, capacity, and resilience to promote caring and instructional equity in low-income areas and urban schools. The science teacher education program focuses on student populations traditionally under-served providing high-quality secondary-science programs, especially to students of color and students whose first language is not English.

The science education program is guided by the mission to reform urban schooling. It seeks to demonstrate that schools for low-income children of color can become rich, rigorous, socially just, and caring learning communities where all children succeed. Its principles include (a) a social justice agenda, that is, in a multicultural society based on pluralism, all cultures and people have equal value; (b) professional education as a "cradle-to-grave" endeavor; (c) collaborations across institutions and communities; (d) professional education, school reform and the university's role in K-12 schooling; (e) a focus on theory and practice; (f) educators' and students' needs for in-depth content knowledge, useful pedagogies, and school cultures that enable serious and sustained engagement in teaching and learning; (g) self-renewal; and (h) diverse, caring, socially responsible learning communities.

There are two primary components of the science education program. The first is a core curriculum that integrates research-based methods with classroom practice by providing advanced study in such areas as cultural foundations, instructional decision-making, and curriculum development. The second is the credential course sequence that guides students towards development of instructional strategies and pedagogical skills for teaching students from diverse cultural and language backgrounds. Sites in local urban school districts provide preservice science teachers with varied sets of student teaching experiences. The science education program is also committed to setting a standard for innovation in classroom computer applications. Computer technology is incorporated into both field-based activities and course work. The primary goal of the program is to prepare teachers to assume a leadership role in urban schools. A demonstrated interest in teaching in a multicultural, urban school environment is of foremost importance when considering an applicant's admission to the program. Year I begins with an academic sequence comprised of the basic curricular requirements for obtaining a master's degree and teaching credentials. Students also participate during the first year in novice ("student") teaching at urban schools. During Year II, students are required to complete a comprehensive context-dependent final project while simultaneously completing their final program course work and Portfolio Presentation.

University at Woodside

The University at Woodside is a large institution (21,000 undergraduate, 8,000 graduate students) in the southeastern United States, about 70 miles from a major metropolitan area noted for its diverse population. The university shares a history with most other southern schools of segregated education and the school's student population still remains anomalous. This university resides in a state which is ranked 11th in population, 55% of the people live in urban areas, and 28% of the state's population is Black (Georgia State Data and Research Center, 1998). Yet only 7% of the entire student population (including international students) is composed of people not representing European-American ethnicity. Its College of Education is one of the largest in the United States with approximately 2,500 undergraduates and 1,900 graduate students. Several years ago, the faculty members adopted a comprehensive multicultural mission statement to promote the development of programs and practices in preparing its students to work and teach all students in the state. In addition, it states that multicultural education is a field of inquiry devoted to research and development of educational policies and practices. The mission statement identifies four areas: instruction, research, service, and administration through which the goals can be addressed. With respect to instruction, four dimensions have been articulated for possible action including (a) the development of curricula; (b) the education of preservice teachers knowledgeable about diverse ways of knowing and learning and competent to implement appropriate teaching strategies to instruct all students; (c) provision of opportunities to reflect on diverse fieldwork and school experiences; and (d) encouragement for preservice teachers to understand and partake in cross-cultural communications.

Several years ago, its preservice secondary science teacher education program underwent a change with the employment of two females who reformed the program by creating a program with goals and infusing multicultural education into all of its courses. The preservice program include the following goals: (a) implementation of a research-based rationale for science teacher education with a five-strand (teacher as discipline liaison, communicator, ethical decision maker, reflective practitioner, and multicultural educator) and six-theme (learning, curriculum, planning, instruction, evaluation, and value/belief) matrix; (b) exploration of what it means to understand science and to investigate how some learners may come to understand these topics; (c) continued thinking about some fundamentally important issues such as "what is science", "what counts as knowledge and as evidence of understanding?", and whose needs or concerns are most important in a classroom; (d) socialization of preservice science teachers into the profession; (e) levels of consciousness that teachers must develop; and (f) challenging preservice science teachers to think on different levels about learning and teaching.

Since the program has a multicultural education strand, we question "How is this strand infused into the program themes?" One departmental decision was to leave the inclusion of multicultural education at the discretion of faculty members, who have the flexibility to include or exclude multiculturalism. Currently, two groups of faculty members have emerged. One group infuses critical multiculturalism throughout program courses taught, whereas the other discusses diverse learning styles in which multiculturalism is treated as an isolated short topic.

DISCUSSION

Elements of the Science Teacher Education Programs

There are commonalties that tend to be representative of all three programs that are noteworthy. First, among the three institutions, there is an awareness of and commitment to addressing diversity through systemic and programmatic initiatives. However, redesigning or restructuring a university's or department's policies, programs, and courses, faculty academic rights remain paramount to any discourse, decision, or implementation. Second, each university is situated in a location that offers collaborative opportunities for school and community partnerships. All three programs have educational linkages that extend into school settings and serve students from low-income families. They strongly emphasize the need for preservice science teachers to develop, organize, and instruct students so that science is a subject for all learners. An example of this can be seen at Oceanside, where there is a center for urban studies based on the idea to strengthen and encourage more parental involvement.

Third, in varying degrees, each program has required preservice science education teachers to enroll in academic coursework that infuses multicultural education content and opportunities to construct, implement, and evaluate reflective practice. Through such efforts, there is a goal that preservice science teachers will be able to demonstrate a more critical multicultural understanding of themselves, their teaching, and their students. In this context, university faculty and students need to challenge their preexisting assumptions, beliefs, and ideologies with respect to science, multiculturalism, and science education (Ogawa, 1998; Rodriguez, 1997, 1998). In part, this was evidenced in how the Native American preservice science teacher at the University at Riverside was able to report to school officials their error in misidentifying students' identities and supposed lack of motivation to learn. This preservice science teacher demonstrated her use of critical multicultural knowledge by seeking and highlighting the motivating factors that enabled her to successfully teach in a more student-centered learning environment. Lastly, all three programs stand as benchmarks for leading institutions in their pursuit towards effectively restructuring science teacher education programs. However, for the most part, the three programs have not yet attained critical multicultural science education programs.

There are several programmatic features of the three science teacher education programs that distinguish them. One feature is the length of time each has been involved in designing and restructuring their programs. This is likely to be result of the demographic changes that have impacted the state within which these programs reside. Certainly, the University at Riverside and Oceanside are at the leading edge since these two sites are major national and international hubs that draw people and students from across the world. The University at Woodside, however, while financially competitive with other university systems at the national level, tends not to be as geographically ideal for international and non-state residents. Instead, the institution has served primarily the state's population of students and as a consequent of its very recent involvement to address issues of diversity, change has occurred more slowly. A second distinction is the amount of attention, time, and funding given in restructuring their science-teacher education programs. The University at Riverside and Oceanside received substantial amounts of money to initiate and implement their programs. Significant events include the establishment of a center at Oceanside, the involvement of the Board of Regents at Riverside, and requirement of a multicultural course or some corresponding experience at Woodside.

The programs differ as a consequence of the settings. For example, the University at Oceanside has an entire program promulgating not only how unique, important, and comprehensive a multicultural science education program needs to be, but also a program that appears to critically examine current research, assumptions, and variables that define multiculturalism. In contrast, although change at the University at Riverside is comprehensive and pragmatic, at present, it is engaged in implementation and does not have a research program devoted to critical multicultural issues. Finally, the University at Woodside attempts to define the essential elements required for a science teacher to be effective and capable of meeting the needs of all learners. In the coming years, it will be interesting to follow the extent to which these three approaches best serve the long-term needs, interests, and requirements for producing successful multicultural science teachers.

Attributes of a Critical Multicultural Science Teacher Education Program

In this section, we describe important attributes of successful development and implementation of a critical multicultural secondary-science teacher-education program (see Table 1). First, a stated theoretical program framework must be adapted or developed such that goals and objectives are infused throughout the program and are sufficiently operationalized to allow evaluation of their effectiveness. Guidance for a framework may be derived and synthesized from numerous works (Atwater, 1993; Banks, 1988; Baptiste, et. al., 1980; Grant, 1991; Sleeter & McLaren, 1995). Among the three secondary science teacher education programs discussed, the University at Oceanside has a well-defined theoretical framework of critical multiculturalism. Second, institutional systemic commitment is required. Science teacher education programs "inherit" their future professionals from the science disciplines. Our future educators in science must have opportunities to question or probe their stereotypes, prejudices, misconceptions, and any discriminating actions throughout their college experience. For reforms to be sustained, systems must be changed. At two of our universities, University at Oceanside and University at Riverside top administrators, along with deans and faculty members, are committed to some kind of multiculturalism.

Table 1. Major Attributes and States of the Three Multicultural Programs Described

	Institution and Program		
Program Attribute	*Oceanside*	*Riverside*	*Woodside*
Level of theoretical Framework	Critical multiculturalism	Multiculturalism	Multiculturalism
Level of inclusion and department	Systemic	Systemic	Partial college
Active teacher recruitment	Yes	Yes	?[1]
Relevant coursework	Yes	Yes	Partial
Relevant field experiences	Yes	Yes	Partial
Opportunities for student reflection	Yes	Yes	?
Research agenda	Yes	Partial	?
Documentation of program effectiveness	Yes	?	?

[1] *No known supporting documentation*

Third, the preservice science-teacher education program should include active recruitment of students who continue to be underrepresented in the sciences. Students need to be able to work with dedicated and knowledgeable mentors. Fourth, relevant course work, fieldwork, and opportunities for student reflection need to be integrated throughout the program for systemic change to be maintained (e.g., Rodriguez, 1998). Again, two out of the three programs discussed in the chapter possess these attributes. Graduates of programs either have the knowledge and skills to teach all students or they do not have the knowledge and skills. Fifth, the science education department needs to create a research agenda to document its program's effectiveness. Few science teacher education programs conduct research on their own programs to determine their program effectiveness. One of the goals of any program should be the evaluation of its effectiveness. Only one program has included in its goal a research agenda for its effectiveness. The University at Riverside program does have a research agenda, but it is not articulated in its goals. The three secondary science education programs previously described reflect different states of development for its theoretical framework, level of inclusion, preservice teacher program infusion, research agenda, and documentation of program effectiveness.

Assessment and Evaluation of Science Teacher Education Programs

What are the assessment and evaluation criteria to be employed to determine the effectiveness and success of a secondary science-teacher education program that embraces critical multiculturalism? First, at the state level prospective teachers are usually given credentials and licenses to help students' dreams to become a reality. The responsibility then rests upon the state's policy that teachers are prepared to teach all of a state's children with the assumption that all students can learn science (Atwater & Brown, 1999). Second, there must exist a systemic and agreed-upon philosophy with short and long term goals at the university or college level. In a democratic society, goals should be dynamic, responsive, and proactive. Third, several criteria may be utilized to evaluate a program's success at the department level. Critical multicultural science-teacher education programs are engaged in producing future teachers who reflect a state's population. They provide opportunities for integration and reflection of theory and practice during which time prospective teachers reflect on their personal epistemologies and beliefs about teaching and learning. Finally, they implement curricula that reflect two important attributes of science, knowing science and ways of knowing science.

Fourth, several professional science teaching associations have adopted either position statements or standards such as the National Science Teachers Association (1991, 1999) and Association for Multicultural Science Education. Accordingly, science teachers should have the knowledge and skills so that all of their students are scientifically literate, have positive self-concept, possess the knowledge and skills to be successful participants in a democratic society, and knowledgeable about career opportunities in science related fields. In addition, science teachers must be able to assist their students in crossing the cultural borders between their own culture and secondary science so that they are empowered to employ their understanding of natural phenomena to make their worlds better for themselves and others (Aikenhead, 1996; Atwater & Brown, 1999). If prospective teachers can provide opportunities for students to utilize their science knowledge and skills to change their worlds, then this is an indicator of a successful critical multicultural science teacher education program. Last, it would be hypocritical to posit that there exists only one way to structure a secondary science teacher education program. Critical multiculturalism depends upon the beauty of pluralism. This is also true about assessment and evaluation of teacher education programs.

Summary

This section deals with multicultural education and critical multiculturalism and analyzes three secondary science teacher education programs at large research universities in the United States to determine how these programs implemented multicultural education. They possess both similar and distinct elements. All three of the programs posses a theoretical framework, some level of support from administration, have relevant course work that promotes multiculturalism and diversity, and provide opportunities for prospective secondary teachers to reflect upon both equitable and inequitable educational policies and practices, and have a research agenda. The difference in these three programs is the degree to which they have these components. Only one program has critical multiculturalism as a theoretical framework and documents the effectiveness of its secondary science teacher education program. Two of the programs (a) actively recruit so that their student populations reflects the population of their states and (b) provide opportunities for students to reflect upon both equitable and inequitable educational policies and practices. In order for secondary-science teacher education programs to become critical multicultural, systemic reformation is required. Mission statements guided by critical multiculturalism emphasize the need for multicultural field settings and students. Teacher educators need funding to reform programs, must work with scientists so that preservice teachers understand that science is only one way of understanding natural phenomena, and conduct research in the area of critical multicultural education. A detailed strategy for ascertaining the effectiveness of teacher programs is needed to determine if their graduates have the knowledge and skills to educate students to utilize their science understanding to change their world and the world of others to be more equitable.

IMPLICATIONS

Implications for Future Science Teacher Education Program Development

Change in programs requires systemic transformation at all levels in the schools and colleges with all players participating in proactive and reactive roles. Learning must take place at all levels of the science education structure, that is, it must involve professors, preservice science teachers, curriculum, etc. Teacher educators must be knowledgeable and possess the skills to critically infuse multiculturalism in science teacher education programs. Teacher educators must embrace and pursue life-long learning opportunities. Decisions must be made with regard to the university's mission in preparing students for future roles in the profession. Today, one important role of science education programs is to prepare quality science teachers for multicultural classrooms. Multicultural classrooms are not an urban phenomenon.

Mission statements of universities and departments reflect their conceptual frameworks that ground and guide their programs. If programs are guided by critical multiculturalism then the teacher educators are obligated to have students involved in teaching in a variety of school settings. Mission statements that emphasize multicultural education may provide a non-prescriptive framework for departments in which preservice science teacher education programs reside. The following questions need to be addressed with respect to the mission of the department or programs. Does the mission document embody a political, moral, or ethical agenda? Is the status of multicultural education a recognized field of inquiry/study? Should the responsibility and actions be departmental, programmatic, or individual in nature? What standards of measurement, qualitative and quantitative, can be used to determine effectiveness and goal achievement?

Preservice science teachers must first have access to multicultural science classrooms. However, even when science educators have the appropriate knowledge and skills, there are other necessary conditions for success. For example, Rodriguez (1998) has reported on two types of resistance found in bringing about change in preservice science teachers: "*resistance to ideological change and resistance to pedagogical change*" (p. 589). Rodriguez concludes that strategies are needed to counter the resistance of preservice science teachers to change; these strategies include dialogic conversations, metacognition, reflexivity, trust among the participants, and an authentic classroom environment.

If successful science educators and science teachers are not available to assist their preservice science teachers in development and implementation of critical multicultural perspectives, knowledge, and skills, then the success of secondary science teacher education programs are in jeopardy. After the last field experience and prior to graduation, preservice science teachers must come away with the capability of seeing critical multicultural education as a lens for viewing their teaching philosophy and practices.

Policy makers need to ensure that secondary science teacher education programs and high school science programs have competent teachers who are able to teach all students they find in their classes. The policies must include the following standards: (a) a declared commitment to critical multicultural science teacher education; (b) appropriate funding for preservice science teacher education programs science teaching; (c) education research that enhances science teaching and learning; and (d) monitoring and evaluating science teacher education programs to be at the forefront.

We have not discussed the continued programmatic division between faculties of science and science education. As a consequence, science educators have little control concerning what and how science is taught in courses preceding admission to a teacher education program. After admission to a science education program, understanding science as a way of knowing is one of the problems, but another is more fundamental: knowledge of science/society concepts with dual meanings. For example, the concepts of race and racism (Gould, 1977; Moore, 1999; Willinsky, 1998) are typically not discussed in science and science education courses. Conversely, critical multicultural science education is no longer a question of why or when, but rather, where is this opportunity going to take us.

Implications for Research

Only a few research studies have so far been conducted with a focus on multicultural science teacher education. Hence, many of our practices are based on findings in other disciplines. Since science is viewed very differently than any other subject, researchers must investigate both policies and practices in science teacher education to determine if teacher educators are educating science teachers who can instruct all of the students they find in their classrooms. The answers may vary on the context of the teacher education programs and the career goals of the preservice science teachers (e.g., programs will vary according to size of institution, available funding, and the interest of science education faculty). Such research may provide us with answers to slogans such as what *Science for All* means in the day-to-day praxis of science teaching and science teacher education.

REFERENCES

Aikenhead, G. (1996). Science education: Border crossing into the subculture of science. *Studies in Science Education, 27*, 1-52.

American Association for the Advancement of Science. (1989). *Project 2061: Science for all Americans.* Washington, D.C.: Author.

American Association for the Advancement of Science. (1993). *Benchmarks for science literacy.* New York: Oxford University Press.

Amphrey, W. G. (1997). Some hard-knock lessons about public-private partnerships. *School Administrator, 54,* 10-12.

Anderson, J. A. (1994). Examining teaching styles and student learning styles in science and mathematics classrooms. In M. M. Atwater, K. Radzik-Marsh, & M. Strutchens (Eds.), *Multicultural education: Infusion of all* (pp. 93-106). Athens, GA: College of Education, the University of Georgia.

Atwater, M. M. (1993). Multicultural science education: Perspectives, definitions, and research agenda. *Science Education, 77,* 661-668.

Atwater, M. M. (1996a). Teacher education and multicultural education: Implications for science education research. *Journal of Science Teacher Education, 7,* 1-21.

Atwater, M. M. (1996b). Social constructivism: Infusing into the multicultural science education research agenda. *Journal of Research in Science Teaching, 33,* 821-837.

Atwater, M. M., & Brown, M. L. (1999). Inclusive reform: Including all students in the science education movement. *The Science Teacher, 66*(3), 44-48.

Atwater, M. M., & Riley, J. P. (1993). Multicultural science education: Perspectives, definitions, and research agenda. *Science Education, 77,* 661-668.

Ausubel, D. P., Novak, J. D., Hanesean, H. (1968). *Educational psychology: A cognitive view.* New York: Holt, Rinehart, and Winston, Inc.

Baptiste, H. P., & Baptiste, M. L. (1979). *Developing the multicultural process in classroom instruction: Competencies for teachers, cognitive competencies, Vol. 1.* Washington, DC: University Press of American Inc. (ERIC Document Reproduction Service No. ED 197032).

Baptiste, Jr., H. P., Baptiste, M. L., & Gollnick, D. M. (1980). *Multicultural teacher education: Preparing educators to provide educational equity. Vol. I.* Washington, DC: American Association of Colleges for Teacher Education.

Banks, J. A. (1981). *Multiethnic education: Theory and practice.* Boston: Allyn and Bacon.

Banks, J. A. (1988). *Multicultural education: Issues and perspectives* (2nd ed.). Boston: Allyn and Bacon.

Banks, J. A. (1993). The canon debate, knowledge construction, and multicultural education. *Educational Researcher, 22*(5), 4-14.

Bennett, C. I. (1995). *Comprehensive multicultural education: Theory and practice* (3rd edition). Boston: Allyn and Bacon.

California State Data Center (1998). Department of Finance. [On line] Available: http://www.dof.ca.gov/html/ Demograp/data.htm

Chavez, R. (1996). *Multicultural education in the everyday: A renassaince for the recommitted.* Washington, DC: American Association of Colleges for Teacher Education. (ERIC Document Reproduction Service No. ED 393 799)

Childs, S., & Killick, D. (1993). Indigenous African metallurgy: Nature and culture. *Annual Review of Anthropolog, 22,* 317 - 337.

Cobb, P. (1994). Where is the mind? Constructivist and sociocultural perspectives on mathematical development. *Educational Researcher, 23,* 13-20.

Education Commission of the States (1997). *Emerging urban school district governance models: Policy brief,* (Report No. 1. UE-97-3). Denver, CO: Education Commission of the States Distribution Center. (ERIC Document Reproduction Service No. ED 406 494).

Gay, G. (1995). Mirror images on common issues: Parallels between multicultural education and critical pedagogy. In C. E. Sleeter & P. L. McLaren (Eds.), *Multicultural education, critical pedagogy, and the politics of difference* (pp. 155-190). Albany, NY: State University of New York Press.

Georgia State Data and Research Center. (1998). Georgia Institute of Technology. Available: http://sdrent.pp.gatech.edu/.

Giroux, H. (1985). *Theory and resistance in education: A pedagogy for opposition.* South Hadley, MA: Bergin & Garvey.

Gould, S. (1977). *Ever since Darwin: Reflections in natural history.* New York: W. W. Norton & Co.

Grant, C. (1991). Culture and teaching: What do teachers need to know? In M. Kennedy (Ed.), *Teaching academic subjects to diverse learners*(pp. 237-256). New York: Teachers College Press.

Grant, C. A., & Ladson-Billings, G. (1997). *Dictionary of multicultural education.* Phoenix, AZ: The Oryx Press.

Green, III, P.C. (1997). To a peaceful settlement: Using constructive methods to terminate contracts between private corporations and school districts. *Equity & Excellence in Education, 30,* 39-48.

Guba, E. G., & Lincoln, Y. S. (1994). Competing paradigms in qualitative research. In N. K. Dengin and Y. S. Lincoln (Eds.). *Handbook of qualitative research* (pp. 105- 117). Thousand Oaks, CA: SAGE Publications.

Hess, D. J. (1998). *Science and technology in a multicultural world: The cultural politics of facts and artifacts.* New York: Columbia University Press.

Jegede, O. (1994). African cultural perspectives and the teaching of science. In J. Solomon & G. Aikenhead (Eds.), *STS Education: International perspectives on reform* (pp. 120-130). New York: Teachers College Press.

Kinchloe, J. (1993). *Toward a critical politics of teacher thinking: Mapping the post modern.* Westport, CT: Bergin & Garvey.

Kinchloe, J. (1995). *Toil and trouble: Good work, smart workers, and the integration of academic and vocational education.* New York: Peter Lang.

Kozol, J. (1991). *Savage inequalities.* New York: Crown Publishers.

Ladson-Billings, G. (1994). *The dreamkeepers: Successful teachers of African American children.* San Francisco: Jossey-Bass.

Lather, P. (1991). *Getting smart: Feminist research and pedagogy within the post modern.* New York: Routledge.

Latour, B. (1987). *Science in action: How to follow scientists and engineers through society.* Cambridge, MA: Harvard University Press.

McLaren, P. (1989). *Life in schools: An introduction to critical pedagogy in the foundations of education.* New York: Longman.

McLaren, P. (1995). *Critical pedagogy and predatory culture: Oppositional politics in a post modern era.* New York: Routledge.

Moore, R. (1999). Science, objectivity & racism. *The American Biology Teacher, 61,* 242.

Murfin, B. (1996). An African Chemistry Connection. *The Science Teacher, 63,* 36-39.

National Commission on Teaching and America's Future. (1996). *What matters most: Teaching for America's future.* New York: Author.

National Council for Accreditation of Teacher Education. (1997). *Standards, procedures, and policies for the accreditation of professional education unit.* Washington, DC: Author.

National Research Council. (1990). *Fulfilling the premise: Biology education in the nation's schools.* Washington, DC: National Academy Press.

National Research Council (1996). *National science education standards.* Washington, DC: National Academy Press.

National Science Foundation. (1997). *Directorate for education and human resources* (Report No. 47.076). Washington, DC: Author.

National Science Teachers Association. (1991). An NSTA position statement: Multicultural science education. In *NSTA Handbook.* [On-line]. Available: http://www.nsta.org/handbook/multi.htm.

National Science Teachers Association. (1998). *NSTA standards for science teacher education* [On line]. Available: http://www.iuk.edu/faculty/sgilbert/nstastand98.htm.

National Science Teachers Association. (1999). *An NSTA position statement: The national science education standards: A vision for the improvement of science teaching and learning.* [On-line]. Available: http://www.nsta.org/handbook/nses.htm.

New York State Data Center. (1998). Division of Policy & Research, Department of Economic Development. Available: http://205.232.252.23/nysdc/.

Ogawa, M. (1998). A cultural history of science education in Japan: An epic description. In W. W. Cobern (Ed.), *Socio-cultural perspectives on science education: An international dialogue* (pp. 139-161). Dordrecht: Kluwer Academic Publishers.

Page, R. (1987). Teachers' perceptions of students: A link between classrooms, school cultures, and the social order. *Anthropology and Education Quarterly, 18*, 77-99.

Rodriguez, A. J. (1997). The dangerous discourse of invisibility: A citique of the National Research Council's National Science Education Standards. *Journal of Research in Science Teaching, 34*, 19-37.

Rodriguez, A. J. (1998). Strategies for counterresistance: Toward sociotransformative constructivism and learning to teach science for diversity and for understanding. *Journal of Research in Science Teaching, 35*, 589-622.

Schmidt, P., & Childs, S. (1995). Ancient African iron production. *American Scientists, 83*, 524-533.

Sleeter, C., & McLaren, P. (Eds.). (1995). *Multicultural education, critical pedagogy, and the politics of difference*. Albany, NY: State University of New York Press.

Vygotsky, L. S. (1978). *Mind in society: The development of higher psychological processes.* Cambridge, MA: Harvard University Press.

Wilder, M. A. (1995). Professional development schools: Restructuring teacher education programs and hierarchies. In H. G. Petrie (Ed.), *Professionalization, partnership and power: Building professional development schools* (pp. 253-268). Albany: State University of New York Press.

Willinsky, J. (1998). The obscured and present meaning of race in science education. In D. Roberts & L. Ostman (Eds.), *Problems of meaning in science curriculum* (pp. 73-85). New York: Teachers College Press.

PORTRAITS OF PROFESSIONAL DEVELOPMENT MODELS IN SCIENCE TEACHER EDUCATION: A SYNTHESIS OF PERSPECTIVES AND ISSUES

Thomas R. Koballa Jr. & Deborah J. Tippins
University of Georgia

The professional development models highlighted in this book are not intended to represent "best practices". Indeed, the notion of exemplary models of professional development in science teacher education is somewhat paradoxical. On the one hand, the various models seem to be identified with a norm toward which all science teachers should clearly be moving. On the other hand, it is apparent that differences in professional learning goals and the context in which teacher learning takes place contribute to the diversity across programs.

Drawing on knowledge of past efforts and current research, the models attempt to move beyond the conventional reasons of the traditional paradigm of professional development with its emphasis on theory first-then practice, teacher preparation for immediate concern versus life-long learning and similar "myths". Over the years, these myths, and others like them, have "served as a framework to repress certain notions of pedagogy while others can be facilitative to generating alternative images of teaching and learning" (Britzman, 1991, p. 8). The chapters in this book attempt to situate professional development for science teachers within a new paradigm, one that Loucks-Horsley, Hewson, Love and Stiles (1998) have characterized as a shift in emphasis from "transmission of knowledge to experiential learning, from reliance of existing research findings to examining one's own teaching practice, from individual-focused to collaborative learning, and from mimicking best practice to problem-focused learning" (p. xv).

Professional development in science teacher education carries with it a powerful sociohistorical legacy. This legacy is reflected in the conversations of science educators, who continue to ask the question, why is there a gap between theory and practice? Across the various examples illustrated in this book, we can ascertain the relative places of theory and practice within diverse professional development models. While an examination of the underlying assumptions about the interplay between theory and practice as characterized in these models is beyond the scope of this synthesis, we nevertheless feel compelled to express some thoughts concerning this relationship. Embedded within each of the illustrated models of professional development is a particular worldview that implies a particular kind of relationship between theory and practice. In most cases, this relationship is not implicitly acknowledged but can be inferred by the reader.

Both control and communication models (e.g., King & Young, 1986) are useful in understanding the theory/practice relationship embedded within a particular

D.R. Lavoie and W.-M. Roth (eds.), Models of Science Teacher Preparation, 213–224.

model. Control models, for example, are based on the premise that theory can be put into practice by manipulating certain conditions. This model is inherently problematic when we consider the human dimensions of science teaching and learning. Communication models, as rival descriptions of the relationship between theory and practice, emphasize the use of persuasion to influence less than certain outcomes. The important idea, here, is that any model of professional development which reflects the non-universal values of only one group of society is ideological in nature. The presence of these tacit values has clear implications for the kind of practice toward which the theory tends. Thus, we encourage the readers of this book to look closely not only at the logical structure of the relationship between theory and practice represented in the various chapters, but at the embedded assumptions that represent a particular ideology.

COLLABORATIVE AND APPRENTICESHIP MODELS

The chapters in the first section of the book describe alternative models of science teacher education. The focus of some chapters is preservice education programs, while others look more at the professional development of science teaching veterans. The need to improve upon the traditional practices of science teacher education and the importance of collaboration among the different constituencies that have a stake in science teacher education are themes that cut across chapters.

Aiding to position the models described in the chapters by Roth, by Peterson and Treagust, and by Hargrave, Thompson and Glass in the discourse of teacher education is Northfield's (1998) description of how learning about teaching and school experience are related. Northfield used Farnham-Diggory's (1994) schema for classifying models of science teacher education into three general approaches—behavior, apprenticeship, and development. The behavior approach is exemplified by traditional models of science teacher education where theory is developed through university experiences and then applied in the practical world of the classroom. It is this behavior approach that is rejected by the three models presented by the chapter authors because it tends to contribute to the gap between university teacher education programs and school experiences, in part by discounting the work of teachers and their contributions to teacher education.

As Hargrave, Thompson and Glass's chapter title indicates, the TEAMS model is an example of the apprenticeship approach. The authors describe their efforts to make visible to preservice elementary teachers the understandings, skills and reasoning of scientists and science teachers through cognitive apprenticeship. Critical to successful learning by means of cognitive apprenticeship is "deliberately bringing the thinking of [both the learner and expert] to the surface" so that the cognitive and megacognitive processes used by the expert can be observed, understood and practiced by the learner (Collins, Brown & Holum, 1991, p. 9). By designing learning experiences for preservice teachers that have them working with veteran scientists in their laboratories and science teachers in their classrooms, the TEAMS model operationalizes assumptions central to the apprenticeship approach: education is a process of enculturation and the dynamic and complex understandings associated with science teaching can best be constructed in context. Implicit within this model is a "view of learning to teach science that might be characterized as a type of... authentic task that learners must help each other accomplish" (Tippins, Kemp, & Ogura, 2000, p. 191). This notion of apprenticeship contrasts with the more traditional conception of apprenticeship as acquiring the skills of the master. It characterizes cognitive apprenticeship as a type of shared problem solving and reflection set within an authentic context.

Roth's coteaching model is a clear example of Farnham-Diggory's (1994) development approach. According to this approach, learning is seen as "continually reshaping personal beliefs" (Northfield, 1998, p. 696) and best occurs in the context of practice. In Roth's coteaching model, the learner's personal beliefs about science teaching are developed or reshaped by working "at the elbow" of a master teacher. Central to Roth's developmental approach is an awareness of the learner's biography, including the image of self-as-teacher and developmental awareness of preexisting beliefs and images. The underlying assumption of a developmental approach such as Roth's co-learning model is that the design and context of science teacher education programs should reflect the genuine needs of novices. The concept of knowledgeability, with its focus on split-second practical decision-making, and that of spielraum, which highlights the "sense of heightened awareness" available to the experienced teacher, bring attention to the two modes of learning accessible through coteaching. The first mode is visible to the learner and makes use of reflection. It is this mode of learning that is also the focus of cognitive apprenticeship models. The second mode involves "learning in practice without bringing the practices into awareness" (p. 18) and is not normally taken into account by models of teacher education.

Peterson and Treagust's problem-based learning (PBL) model is not a clear example of either the apprenticeship or development approaches. Rather, it is a hybrid of sorts, combining aspects of both. Like the TEAMS and coteaching models, the PBL model strives to integrate theory and practice by engaging prospective teachers in contextual learning experiences. This is done through the use of classroom cases that address the knowledge bases for teaching and the pedagogical reasoning associated with the application of this knowledge. This approach exemplifies many of the cognitive apprenticeship methods such as coaching, scaffolding, articulation and reflection, as prospective teachers collaboratively analyze their experience. However, it differs from apprenticeship in terms of the experienced teachers' level of involvement in the co-learning process. To guide the development of personal beliefs about science teaching, the prospective teachers rely primarily on their peers, with occasional assistance from teachers.

As is true in the TEAMS model, the focus in the PBL model is on the prospective teachers' conscious knowing. Reflection-on-action is encouraged to increase awareness of the act of teaching, but not addressed is the unconscious learning considered by the coteaching model. By situating learning within a case and/or case discussion, the PBL model creates an opportunity for the practitioner to become a researcher in the context of practice. As such, this approach to problem solving might be located along a continuum of technical-rational and critical-reflective approaches to problem solving. However, as Wade and Moje (1997) suggest, this dichotomy might serve to oversimplify approaches to problem solving as represented in the PBL model. These authors adopt the term "blended voices" to emphasize that "thoughts and utterances are socially constructed and historical, "overpopulated" with the voices of others—reflecting both critical reflective thinking and technical rational thinking" (p. 78).

As alternatives to the behavior approach, the three models share some important features. All attempt to blur the differences between academic knowledge and professional knowledge. They stress the need to shift science teacher education from translating theory into practice to "providing better conditions for learning to teach" (Northfield, 1998, p. 698). They also emphasize the active role of the learner in the process of learning to teach. In no case is learning to teach a passive activity. The need for active learning is accommodated in the models through small group instruction or working one-on-one. The benefits of these personalized learning opportunities are tremendous, but may not be sustainable without external resources. The activities associated with the TEAMS model were supported by an external grant and the coteaching model relied on the expertise provided by university researchers. Only the PBL model provides guidance for scaling up to work with classes of 20–30 students.

Additionally, the three models acknowledge the importance of situated learning (Brown, Collins & Duguid, 1989) by linking the knowledge of teaching to the conditions of its use. This is seen in the centrality of the school experience and the real work of teachers to the coteaching and TEAMS models. While not engaging the preservice teachers in actual school experiences, the PBL model uses cases as tools to have students explore resources and integrate knowledge while solving school-based problems of science teaching and learning. Seeming to reinforce the need to connect knowledge and the conditions of its application is the less than successful attempt by Hargrave and her colleagues to have scientists serve as cognitive mentors for preservice teachers in the TEAMS model.

Of the three models, Roth's model of coteaching is the most appealing to us because of the way in which learning to teach is embedded in the act of teaching and its ability to explain the actions of teachers that are not brought to conscious awareness. However, it is limited by the need for the learner to be partnered with a master teacher who is not only a skilled teacher of students but a skilled teacher of teachers. As our own experiences tell us, persons learning to teach benefit from multiple perspectives. An excellent classroom teacher should be entrusted to provide one perspective about how to teach to help students learn, but this should not be the only perspective. We agree with Roth that coteaching may be a very cost efficient option for inservice teacher development, but we find its use in preservice teacher education more difficult. If choosing to base a teacher education program on this model, much care must be taken when pairing master teachers and novices. Two contextual factors of particular importance to this model include the nature of the personal relationship that develops between the novice and experienced teacher and the extent to which building principals afford teachers a degree of autonomy in the construction of coteaching relationships.

The chapter by Fetters and Vellom uses the Holmes Group model of Professional Development School as a backdrop for discussing issues and concerns related to the preparation and support of novice teachers. The stages that a school-university partnership might go through as it develops are presented as four vignettes. The first two vignettes highlight the growing involvement of university faculty as a school wrestles with how to improve low student scores on a state-wide science test while responding to a request to work with preservice science teachers. The final two focus on the changing roles of university professors and school teachers and administrators as a school-university partnership matures. Through each vignette, the authors identify the numerous barriers likely to derail the collaboration and provide recommendations to overcome them. While the focus is squarely on how university-school partnerships can make a difference in preservice teacher education, attention is also paid to the other two issues central to the Professional Development School model—inservice professional development and inquiry into teaching and learning. Not addressed in the chapter are the contributions of scientists and other content experts as well as business, industry and community leaders to realize the Holmes Group vision of a professional development school becoming the center of a learning community.

The chapter by Barufaldi and Reinhartz broadens the discussion of collaboration beyond school-university partnerships. Their focus is on a statewide professional development program and the collaborative dimensions critical to its success. No doubt the organizational complexity of the program—involving 20 regional collaboratives—brought challenges to the process very different from those experienced in school-university partnerships. Unfortunately, the chapter does not share the false starts, momentary successes and frustrations that surely accompanied the evolution and refinement of the postulates presented for successful statewide collaboration. From our perspective, this would have made for interesting reading. As portrayed in this chapter, the authors' description of a successful statewide collaboration reflects a structuralist approach to reform. It fails to capture the complexity and human dimensions that are integral parts of the process of collaboration. As Sirotnik (1988) explained, "school-university partnerships are not...controlled social experiments... (they are) evolving social experiments by people... struggling with alternative ideas and organizational arrangements and activities for promoting collaboration between typically non-collaborative institutions" (p. 169). Nevertheless, the authors do present their understandings gained from the process as a model from which others can draw insights about organizing professional development for science teachers that is in alignment with the current school reform movement.

While distinctive in context and scope, the two collaboratives described in these chapters are examples of educational reform networks. In common with the sixteen educational reform networks studied by Lieberman & Grolnick (1996, p. 7), the Springvale PSD and Texas Regional Collaboratives promote and foster:

- agendas that are more challenging than prescriptive;
- learning that is more indirect than direct;
- formats that are more collaborative than individualistic;
- work that is intentionally more integrated than fragmented;
- leadership more facilitative than directive;
- thinking that encourages multiple perspectives;
- values that are both content-specific and generalized; and
- structures more movement-like than organization-like.

It is these shared themes that distinguish the two collaboratives as alternative organizational structures for science teacher education.

Our look across the models described in these chapters highlights how important understandings of learning are to the development of science teacher education programs. The rejection of the notion that teacher knowledge is an additive quantity that is acquired in one place (the university) and applied in another (the school) has led to new ways of conceptualizing teacher learning. Viewing teacher learning as a process of acculturation or continual reshaping of personal beliefs has implication for the development of teacher education programs as seen in the models described by Roth, by Peterson and Treagust, and by Hargrave and Thompson. Considering teacher learning in the context of school reform adds complexity to the process of teacher education. This complexity is manifested in the need for "school leaders and teachers to understand what schools are for and why reform is needed" (Kincheloe, Steinberg, & Tippins, 1999, p. 187). This requires a "space" for joining the community and school in a partnership for mutual conversations. It also poses a challenge for teachers as they consider how best to integrate multiple forms of community knowledge into science classrooms. As the chapters by Fetters and Vellom and by Barufaldi and Reinhartz point out, building learning communities, whether involving one site or many, is a challenging task. The needs and concerns of all members of the learning community must be successfully negotiated in creating and sustaining a partnership or network. There are not simple, straight forward answers to the complex problems associated with the teacher education and its influence on student learning (Darling-Hammond, 1998; Lieberman, 1995; Lieberman & Grolnick, 1996), but the models presented in the chapters of this section move us forward in our search for solutions.

SPECIAL-ISSUES DRIVEN MODELS

It is clear that students' cultural backgrounds influence their science learning. If all students are to become successful science learners then teachers of science need to be responsive to the learning needs of diverse students. The key to responsive science teaching is the education of teachers, and how to prepare teachers to address the science learning needs of students from diverse backgrounds is a question in search of multiple solutions. The three chapters in this section bring different perspectives to issues of educating teachers to work effectively with diverse science learners.

The slogan of "science for all" has brought attention to the assumptions underlying science teaching and learning for diverse populations. Debates associated with promoting "science for all" have been raised between those who pose science education as a means to enhance citizens' quality of life (Cobern & Loving, 1998) and others who perceive science education as an ideological threat to cultural knowledge. The philosophical backdrop for the arguments presented in these chapters is the ongoing debate among science education researchers about the nature of science as a unique way of knowing. The extreme positions in this debate are those often associated with the positivist and postmodern camps. The differences between these camps are most pronounced in terms of (a) the influence of human subjectivity on research and science and (b) the role of data and evidence in the formulation of explanations, particularly whether explanations should be empirically based or "valid because of community respect, cultural status, and psychological significance" (Loving, 1997, p. 422). Positivists tend to view science as a universal form of knowledge that transcends cultural interpretation, and stress the importance of empirical evidence while acknowledging that there are multiple approaches to scientific investigation (Matthews, 1994). On the other hand, postmodernists focus attention on the influence of culture on science and science learning, and claim that Western science (that taught in schools) is only one among many scientific frameworks for investigating phenomena of the natural world (Loving, 1995). Social critical theory is often used as an epistemological framework by postmodernists to investigate science and science education because of its focus on the ideals of social justice.

The chapters in this section do not proffer the extreme positions associated with either camp, but do draw from them in arguing for different ways to prepare teachers of science to teach a rapidly increasing diverse student population. Social critical theory is discussed as an appropriate organizing paradigm for science teacher education in the chapter by Thomson, Wilder and Atwater and in the one by Rennie. Thomson and his colleagues link a social critical perspective with multicultural education in what they call critical multicultural science education. At the heart of this framework for guiding science teacher education, according to these authors, is the need for preservice teachers to challenge the social forces surrounding, race, class, gender and power relations that shape and influence science education. Their comparison of three secondary science teacher education programs highlight the attributes that likely contribute to an institution's success in implementing a program based on the beliefs of critical multicultural science education.

In her chapter that focuses on gender equity, Rennie makes the point that prospective teachers hold beliefs about science and gender that may not be well-formed and may not be ready to address issues of science teaching and learning from a social critical perspective. Drawing on the work of Willis (1996) in mathematics education, she presents three perspectives—remedial, non-discriminatory and inclusive—in addition to social critical for science teacher educators to consider when striving to prepare prospective teachers to teach for gender equity. The framework, because it presents a continuum of sorts regarding basic assumptions about gender equity associated with science teaching and learning, can be a useful tool for science teacher educators. It could be used in science teacher education classes to help prospective science teachers position themselves with respect to their beliefs about gender equity and to guide their growth as science teaching professionals. If expanded to address assumptions about cultural and racial equity, the framework may serve to lessen the resistance to ideological change among prospective science teachers like that reported by Rodriguez (1998).

Lee and Fradd's chapter does not speak directly to science teacher education, but presents a model for educating students from diverse cultures. Their model of instructional congruence strives to balance the strengths that both positivism and postmodernism can bring to the science education of linguistically diverse learners. This balance is achieved, the authors claim, by simultaneously promoting pupils' learning of science and literacy. By focusing both on science and literacy, the model is designed to help learners "negotiate the cultural boarders" (Cobern & Aikenhead, 1998, p. 42) associated with learning science and English by linking learning in these areas to the pupils' indigenous language and culture. Building on the work of Costa (1995), Aikenhead describes cultural border crossing as a "shift from being one person in one context to being another person in a different setting" (Aikenhead, 1996, p. 273). In other words, for some students transitioning between their worlds in and out of school involves little risk; whereas, other students may experience feelings ranging from a sense of uneasiness to alienation.

In their chapter Lee and Fradd suggest that the hazards of cultural border crossing might be reduced by explicitly promoting the learning of both science and literacy. What seems to be needed in their framework is a questioning of the ideology of school science. What will students gain from negotiating borders into the culture of school science? What does school science mean to these learners in contrast to science situated within local communities? It would be useful to extend the border metaphor at the heart of this model to reflect the doing of science "within the border" of local communities. Lee and Fradd's examples of how the model can be operationalized are drawn from elementary classrooms where the teachers and pupils share the same language and cultural background. These examples that highlight knowing science, doing science, talking science, and scientific habits of mind reveal the key role of the teacher in the effective implementation of the model and raise questions about the usefulness of the model when linguistic and cultural backgrounds of teacher and learners do not match. In contrast to the context for implementing the model described by Lee and Fradd, secondary science teachers are typically knowledgeable of science and science instruction but possess limited knowledge of the cultural backgrounds and languages of their diverse students. What can be done as part of preservice or inservice education to help science teachers establish instructional congruence in their classrooms when they do not share their students' linguistic and cultural backgrounds?

Collectively, the chapters in this section focus the discussion of equity in science education not on the learner but the system that serves the learner. All advocate that the system, which includes science teacher education, is where attention should be directed if all learners are to be successful in science. A message also conveyed by the chapters is that issues of equity must be integrated throughout all initial teacher education coursework and professional development experiences. Important considerations in the education of today's science teachers is helping them become familiar with their students' home language and culture and to contextualize their science teaching to better serve all students. Not touched on directly in the chapters is the influence of poverty on equity in science education, which Barba and Reynolds (1998, p. 925) identify as the "single variable that holds most children back... regardless of race or ethnic group". There is little doubt that inequalities in education mirror inequalities in social class. Yet, as Newman (1993) suggests, a consideration of poverty and its relationship to science education is often masked by discussions of comparable factors such as race, religion or gender. The issue is further complicated in our attempts to answer "who is poor?" As Newman explains, "almost 70 percent of all poor people are white. The reason is almost 80 percent of the population is white, and even though whites have a lower incidence of poverty than the total population, whites still constitute the largest group of poor Americans" (1993, p. 186). This pattern will undoubtedly change as classrooms of the future become complex microcosms of cultural diversity. Future research in science teacher education may wish to address how science teachers can be prepared to better address the science learning needs of students living in poverty.

Important aspects of the current reform in science teacher education involve reforming both instruction and assessment. Aided by the contributions of researchers in the areas of cognitive psychology and constructivism, learning to teach is not longer viewed as the acquisition of content knowledge and classroom survival skills but as a life-long process of professional development. As this process unfolds science teachers develop understandings about many aspects of their work in schools, including instructional strategies and methods that can be used to help students learn science and to keep them safe while doing so. An expanded view of assessment also makes it an integral part of the teacher learning process. Assessment is not only used to evaluate a teacher candidate's competence for licensure, but is applied throughout the teacher education process to support opportunities for learning about teaching. The models presented in the chapters of this section are based on these expanded views of instruction and assessment. They offer the potential to make teacher learning and assessment a seamless process and to significantly improve the science education provided all students.

The capstone experience for students in Wayne State University's initial teacher preparation program is the creation and presentation of a portfolio. Stein's chapter describes the WSU portfolio process from its beginnings through almost eight years of implementation. The story of this model program highlights that role of the portfolio as an evaluative measure and as learning tool and describes the collaborative work among faculty responsible for teacher education at the elementary and secondary levels.

The success of the program is directly tied to the faculty's initial efforts to identify what graduates of WSU teacher education program should know and be able to do. The outcome of this work is articulated in the form of competencies that are shared with students early during the program. These competencies serve as a road map of sorts for students, telling them where they should be at the conclusion of their educational journey. In the WSU program, course instructors and portfolios developed by program graduates offer suggestions about what artifacts students might gather to document their journey. Even with such guidance, considerable thought is required by WSU students to create portfolios that are more than a collection of artifacts. As Stein and others (Dana & Tippins, 1998; Gitomer & Duschl, 1995) point out, a collection of artifacts alone does not document professional growth. Students must explain the significance of the evidence included in their portfolios. When they wrestle with linking videotapes of lessons, personal philosophy statements and pupil work samples to their own professional development, beginning teachers no longer view the portfolio as a program requirement by as something they own (Tellez, 1998).

The dual role of the portfolio process as an evaluative measure and as a learning tool brings with it a certain tension not fully explored in Stein's chapter. In reporting on the uses of portfolios by their students at University of California at Santa Barbara, Jon Snyder and his colleagues (1998) describe how externally imposed criteria associated with state licensure requirements led preservice teachers to view portfolios not as an opportunity for personal reflection and growth but solely as a collection of artifacts for measuring teaching competencies. Their program's solution to this problem was to have students construct two different portfolios—credential and personal inquiry—using the same collection of artifacts. While having students construct two portfolios may not be seen as a viable option for all science teacher education programs, this solution does bring attention to the need to ensure that continual professional growth remains central to the portfolios development process. To maintain this focus, the WSU model uses portfolios as one of several measures of performance and allows for portfolios judged as weak to be improved based on feedback received from the review team and resubmitted. This approach ensures that the portfolio is not viewed as an isolated evaluation instrument. Rather, the WSU model, in attempting to document and evaluate the many experiences, tasks, struggles, growth, and accomplishments of students exiting their science teacher education program, emphasizes the complex nature of this task.

The central message that Lavoie conveys in his chapter is that opportunities to engage in discussion and debate with fellow learners are necessary for meaningful teacher learning. He describes how on-line discussions, organized around driving questions or issues, can involve beginning and veteran teachers in conversations about their own classroom practices and help them develop confidence in using e-mail and other forms of telecommunications technology. An example of the power of on-line discussion is Abell's (2000) case in which she describes how a teacher worked through issues of designing problems-based lessons for her middle grades students with the on-line support of colleagues who teach at schools some distance away. As Lavoie points out, on-line discussions "facilitate the social construction of knowledge" about science teaching and learning. Telecommunications can create new opportunities for sharing information, brainstorming and asking questions of other science teachers.

The ethic of cooperation brought about by expansion of shared information can be empowering for teachers and can change the very nature of their discourse. As Kincheloe, Steinberg & Tippins (1999, p. 182) point out, "teachers need to talk about teaching. Talk about students, "work and parents is not critical discourse—talk about teaching is". Thus as teachers and students increasingly make use of communications technology to enter into conversations about teaching, knowledge becomes kinetic, expanding in multiple directions. Lavoie argues that all science teachers should be introduced to the vast array of learning opportunities that the growing availability of telecommunications technology make possible. Recent developments and innovations in communications technology create new possibilities for building learning networks between schools and communities, and in the process, expand our understanding of what it means to build a science teaching and learning community. Amidst the proliferation of telecommunications technology, science educators must take care to ensure that the application of new technologies is not decontextualized from the critical examination of technology in our lives.

The chapter by Yager, Enger and Guilbert describe features of the secondary-science teacher education programs at the University of Iowa. Constructivism, in different iterations (socio-cultural constructivism, emancipatory constructivism, interactive-constructivist philosophy) serves as the glue that ties the different facets of the program together. Unique to the Iowa program is its alignment with the tenets of the Scope, Sequence and Coordination project. The influence of this alignment on prospective teachers' understandings and actions relative to learner motivation and autonomy in decision-making as well as assessment are described in the two cases of Steve and Terry. Teaching and learning as reflective processes and the need for prospective teachers to engage in real work are two themes that bring coherence to the Iowa program and link key program activities. The strength of the chapter by Yager and his colleagues are the insights that they provide about how elements of conceptual frameworks may be translated into actual program experiences.

A multitude of sources inform the development and activities of science teacher education programs (Coble & Koballa, 1996), but based on the chapter authors' descriptions none appear to be more important than contemporary conceptions of learning. Constructivist views of learning and the application of this knowledge to teaching have had a profound effect on science teacher education. But pertinent understandings about teacher reasoning, pedagogical content knowledge, human development, multiculturalism, and cognitive instruction must not be ignored as departure points for learning to teach science (Howey, 1996). An important message that may be taken from these chapters is that when conceptual frameworks and central themes are articulated, students will be less likely to view their science teacher education as a collection of disparate and disconnected ideas and practices.

REFERENCES

Abell, S. K. (2000). Be prepared! In T. R. Koballa & D. J. Tippins (eds.), *Cases in middle and secondary science education: The promise and dilemmas* (pp. 75-78). Upper Saddle River, NJ: Merrill.

Aikenhead, G. S. (1996). Science education: Border crossing into the subculture of science. *Studies in Science Education, 27*, 1-52.

Barba, R. H., & Reynolds, K. E. (1998). Toward an equitable learning environment in science for Hispanic students. In B. J. Fraser & K. G. Tobin (Eds.), *International handbook of science education* (pp. 925-939). Dordrecht, The Netherlands: Kluwer Academic Press.

Britzman, D. P. (1991). *Practice makes practice*. Albany, NY: State University of New York Press.

Brown, J. S., Collins, A., & Duguid, S. (1989). Situated cognition and the culture of learning. *Educational Researcher, 18*, 32-41.

Cobern , W. W., & Aikenhead, G. S. (1998). Cultural aspects of learning science. In B. J. Fraser & K. G. Tobin (Eds.), *International handbook of science education* (pp. 39-52). Dordrecht, The Netherlands: Kluwer Academic Press.

Coble, C. R., & Koballa, T. R. (1996). Science education. In J. Sikula (ed.), *Handbook of research on teacher education* (pp. 459-484). New York: Macmillan.

Cobern, W. W., & Loving, C.C. (1998, April). *Defining "science" in a multicultural world: Implications for science education*. Paper presented at the Annual Meeting of the National Association for Research in Science Teaching, San Diego, CA.

Collins, A., Brown, J. S., & Holum, A. (1991). Cognitive apprenticeship: Making things visible. *American Educator, 15*, 6-11 & 38-46.

Costa, V. B. (1995). When science is "another world": Relationships between worlds of family, friends, school and science. *Science Education, 79*, 313-333.

Dana, T., & Tippins, D. J. (1998). Portfolios, reflection, and educating prospective teachers of science. In K. Tobin & B. Fraser (Eds.), *International handbook of science education* (pp. 719-732). Dordrecht, The Netherlands: Kluwer.

Darling-Hammond, L. (1998). Teacher learning that supports student learning. *Educational Leadership, 55*, 6-11.

Farnham-Diggory, S. (1994). Paradigms of knowledge and instruction. *Review of Educational Research, 64*, 463-477.

Gitomer, D. H., & Duschl, R. A. (1995). Moving toward a portfolio culture in science education. In S. Glynn & R. Duit (Eds.), *Learning science in schools: Research reforming practice* (pp. 299-326). Mahwah, NJ: Erlbaum.

Howey, K. (1996). Designing coherent and effective teacher education programs. In J. Sikula (ed.), *Handbook of research on teacher education* (pp. 143-170). New York: Macmillan.

Kincheloe, J. L., Steinberg, S. R., & Tippins, D. J. (1999). *The stigma of genius: Einstein, consciousness, and education.* New York: Peter Lang.

King, R. J. R., & Young, R. E. (1986). *A systematic sociology of Australian education.* Sydney: George Allen & Unwin.

Liberman, A. (1995). Practices that support teacher development: Transforming conceptions of teacher learning. *Phi Delta Kappan, 76*, 591-596.

Lieberman, A., & Grolnick, M. (1996). Networks and reform in American education. *Teachers College Record, 98*, 7-45.

Loucks-Horsley, S., Hewson, P. W., Love, N., & Stiles, K. E. (1998). *Designing professional development for teachers of science and mathematics.* Thousand Oaks, CA: Corwin Press, Inc.

Loving, C. C. (1995). Comments on "Multiculturalism, universalism, and science education". *Science Education, 79*, 341-348.

Loving, C. C. (1997). From the summit of truth to its slippery slopes: Science education's journey through positivist-postmodern territory. *American Educational Research Journal, 34*, 421-452.

Matthews, M. R. (1994). *Science teaching: The role of history and philosophy of science.* New York: Routledge.

National Research Council. (1996). *National science education standards.* Washington, D. C.: National Academy Press.

Newman, J. W. (1993). Socioeonomic class and education: In what ways does class affect the educational process? In J. L. Kincheloe & S. R. Steinberg (Eds.), *Thirteen questions: Reframing education's conversations* (p. 186). New York: Peter Lang.

Northfield, J. (1998). Teacher educators and the practice of science teacher education. In B. J. Fraser & K. G. Tobin (eds.), *International handbook of science education* (pp. 695-706). Dordrecht, The Netherlands: Kluwer.

Rodriguez, A. J. (1998). Strategies for counterresistance: Toward sociotransformive constructivism and learning to teach science for diversity and for understanding. *Journal of Research in Science Teaching, 35*, 589-622.

Snyder, J., Lippincott, A., & Bower, D. (1998). The inherent tensions in the multiple uses of portfolios in teacher education. *Teacher Education Quarterly*, Winter, 45-60.

Tellez, K. (1998). Authentic assessment. In J. Sikula (ed.), *Handbook of research on teacher education* (pp. 704-721). New York: Macmillan.

Tippins, D. J., Kemp, A., & Ogura, Y. (2000). Learning to teach science: The curriculum of student teaching in Hiroshima "attached" schools. *Journal of Science Teacher Education, 11*, 189-206.

Wade, S. E. & Moje, E. B. (1997, March). *Verbal interaction patterns in case discussions associated with critical/reflective and technical/rational thinking.* Paper presented at the Annual Meeting of the American Educational Research Association, Chicago.

Willis, S. (1996). Gender justice and the mathematics curriculum: Four perspectives. In L. H. Parker, L. J. Rennie, & B. J. Fraser (Eds.), *Gender, mathematics and science: Shortening the shadow* (pp. 41-51). Dordrecht, The Netherlands: Kluwer Academic Press.

SUBJECT INDEX

Science & Technology Education Library

Series editor: Ken Tobin, *University of Pennsylvania, Philadelphia, USA*

Publications

1. W.-M. Roth: *Authentic School Science.* Knowing and Learning in Open-Inquiry Science Laboratories. 1995 ISBN 0-7923-3088-9; Pb: 0-7923-3307-1
2. L.H. Parker, L.J. Rennie and B.J. Fraser (eds.): *Gender, Science and Mathematics.* Shortening the Shadow. 1996 ISBN 0-7923-3535-X; Pb: 0-7923-3582-1
3. W.-M. Roth: *Designing Communities.* 1997 ISBN 0-7923-4703-X; Pb: 0-7923-4704-8
4. W.W. Cobern (ed.): *Socio-Cultural Perspectives on Science Education.* An International Dialogue. 1998 ISBN 0-7923-4987-3; Pb: 0-7923-4988-1
5. W.F. McComas (ed.): *The Nature of Science in Science Education.* Rationales and Strategies. 1998 ISBN 0-7923-5080-4
6. J. Gess-Newsome and N.C. Lederman (eds.): *Examining Pedagogical Content Knowledge.* The Construct and its Implications for Science Education. 1999 ISBN 0-7923-5903-8
7. J. Wallace and W. Louden: *Teacher's Learning.* Stories of Science Education. 2000 ISBN 0-7923-6259-4; Pb: 0-7923-6260-8
8. D. Shorrocks-Taylor and E.W. Jenkins (eds.): *Learning from Others.* International Comparisons in Education. 2000 ISBN 0-7923-6343-4
9. W.W. Cobern: *Everyday Thoughts about Nature.* A Worldview Investigation of Important Concepts Students Use to Make Sense of Nature with Specific Attention to Science. 2000 ISBN 0-7923-6344-2; Pb: 0-7923-6345-0
10. S.K. Abell (ed.): *Science Teacher Education.* An International Perspective. 2000 ISBN 0-7923-6455-4
11. K.M. Fisher, J.H. Wandersee and D.E. Moody: *Mapping Biology Knowledge.* 2000 ISBN 0-7923-6575-5
12. B. Bell and B. Cowie: *Formative Assessment and Science Education.* 2001 ISBN 0-7923-6768-5; PB: 0-7923-6769-3
13. D.R. Lavoie and W.-M. Roth (eds.): *Models of Science Teacher Preparation.* Theory into Practice. 2001 ISBN 0-7923-7129-1

KLUWER ACADEMIC PUBLISHERS – DORDRECHT / BOSTON / LONDON

MIX
Papier aus verantwortungsvollen Quellen
Paper from responsible sources
FSC® C105338

If you have any concerns about our products,
you can contact us on
ProductSafety@springernature.com

In case Publisher is established outside the EU,
the EU authorized representative is:
Springer Nature Customer Service Center GmbH
Europaplatz 3, 69115 Heidelberg, Germany

Printed by Libri Plureos GmbH
in Hamburg, Germany